Alabama Studio Sewing + Design

A Guide to Hand-Sewing an Alabama Chanin Wardrobe

For Mother, and all the other mothers, grandmothers, sisters, friends, aunts, colleagues, and artisans who have always been there...

Natalie Chanin

Photographs by Rinne Allen, Natalie Chanin, Russ Harrington, Sarah E. Lewis, Robert Rausch, Peter Stanglmayr, and Elizabeth DeRamus
Illustrations by Sun Young Park

Abrams, New York

Contents

Introduction

It's hard to believe that a decade has passed since I returned to my home in Florence, Alabama, to begin the work that led me to create the fashion and lifestyle company Alabama Chanin—and even more surprising that sewing has come to define my life. I didn't do well in home economics class in high school. My first apron fell apart. In design school my attempts to sew on the beautiful Elna machine my mother had bought for me were clumsy at best. Somehow, perhaps as a result of the many hours of my childhood spent sitting beside my grandmothers while they sewed, quilted, crocheted, and embroidered, I emerged as an adult with a passion for sewing and creating. I left home at eighteen, wanting to explore the world and to design. Little did I know that the cotton being grown in my childhood community would eventually lure me back to my family home to sustain the traditions of my grandmothers and to found Alabama Chanin.

I have been asked many times why I choose to write books and, in the process, open-source (that is, freely share) instructions for making Alabama Chanin's couture collections. The answer is not as straightforward as you might think. It is based on my belief that good design should be available to all and my desire to build a company that is sustainable in all of its practices. By sharing our skills in these books, I hope to shed light on not only how we can preserve precious natural resources but also how we can preserve and protect techniques that were once understood as essential survival skills.

While Alabama Chanin dresses, skirts, tops, and coats have been beautifully featured in countless magazines and newspapers, and on television shows and websites, they have also been criticized for being "elitist," and "inaccessible" because of their price. Truth be told, our clothing is extremely expensive. This is because it is made from domestic, organic, custom-dyed cotton jersey that is cut, painted, sewn, and embellished by hand in America by skilled artisans. And while we sell our collection to some of the most upscale stores and clients, we run our business in the most down-to-earth, simple way imaginable. In the beginning, we worked from a three-bedroom, brick, ranch-style house in rural Alabama, a home that my grandfather built. Today we work from a reclaimed textile factory built in the 1980s—when manufacturing was booming in the South. Our studio—which we call The Factory—has become a busy hub, where we concentrate on building a zero-waste company. Our employees earn a living wage, and while none of us is getting rich, at least in terms of our bank accounts, we are, indeed, rich in spirit, belief, passion, and friendship. "Elitist" is the antithesis of how the company works and who I am as a designer, entrepreneur, and citizen.

Our clothing is hand-sewn and elaborately hand-embellished, but the techniques we use—such as straight stitch, appliqué, reverse appliqué, embroidery, and beading—are basic and time-honored. I am committed to the idea of sharing these techniques so that anyone who wants to can use them to enrich their wardrobe, work, and life. When I was approached about writing my first book, *Alabama Stitch Book*, my thought was absolutely *yes*—and while we're at it, let's include instructions and patterns for our two best-selling styles: our corset and our swing skirt. I embraced the idea of providing the instructions and patterns to people who loved our clothing, but simply couldn't afford to buy it, so they could either learn how to make it themselves or hire someone in their community to do it for them. In this way, I could share my work and sustain our hand-sewing traditions with many more people.

Of course, there were doubts from colleagues in the fashion industry who thought that by sharing our "trade secrets" I would be leading the company to ruin. And while I firmly believed that my decision to open-source was the right one, I admit that I was scared when *Alabama Stitch Book* was released in 2008. I questioned if the books would sell and if our couture customers would still want to buy our finished garments now that they could be made by anyone who could take needle to fabric. However, my fears proved groundless when books sold well and interest in our couture collections grew. Readers started to work with the techniques and discovered not only why our garments cost so much but why our garments are *worth* so much. And, at the same time, a completely new field of business opened for us: We began to sell the supplies needed to make our designs (organic cotton jersey, thread, stencils, fabric paint, beads, and project kits) and to host hands-on workshops both in our studio in Florence and around the country.

Alabama Stitch Book led to *Alabama Studio Style*, in which I elaborated on the basic techniques presented in the first book. *Alabama Studio Sewing + Design* elaborates further and archives the variations that can be achieved by manipulating each of the techniques. While each of the three books can be used independently, it is my hope that our readers will be inspired to use them together. The possibilities—and the creative pleasure—are endless when you begin to mix and match techniques on different garments and accessories.

In Chapters 1 through 8 of this book, I present instructions for the stitching and stenciling techniques and garments that are at the core of our collections. In Chapter 9, I show how the techniques and garments can be mixed, matched, and layered to create a complete look. The patterns for all of the garments are printed on a pattern sheet attached at the back of the book. In the Index is a guide to the design choices we made for every garment and fabric shown, so you can re-create them in your own studio. And while I encourage you to do that, I also hope that you will feel inspired to follow another age-old tradition: taking creativity to the next level by adding your own character, voice, and vision.

Alabama Studio Sewing + Design represents a decade of my life, work, and passion. I hope it will inspire you to celebrate the skills of the needleworkers who came before and motivate you to honor them by sustaining their traditions. It is my belief—and experience—that in the process your world, and perhaps even your soul, will be enriched.

Chapter 1
The Basics—Materials + Tools

At the core of all work are the materials and tools that make it possible to do a particular job. At Alabama Chanin, this process starts with pure organic cotton-jersey fabric. Over the last ten years, I've tried to incorporate many different fabrics into our collections; but, while the fabrics and results have been stunning, our customers have always come back asking for cotton jersey.

Cotton-jersey fabric comes in a variety of weights commonly described as ounces per linear yard. For many years we have been using a medium-weight jersey that averages 9.80 ounces per linear yard. We recently added a lightweight jersey that is stretchier than the medium-weight jersey and averages 5.6 ounces per linear yard.

After securing our base fabric, the design process starts with research and inspiration. Over the years I've collected and organized my own library of design, stencil, textile, photography, and other inspirational books that we use as tools to help me create a theme for a project or collection. Next, we choose color palettes for our cotton jersey; develop and prepare stencils; create a library of fabric swatches using techniques and embellishments in Chapters 5–8; and finally translate those fabric swatches into clothing and other projects like the ones in Chapter 9.

Supporting these design, development, and manufacturing processes is a set of good tools. This chapter introduces the materials and tools that we use on a daily basis at Alabama Chanin.

A work table in our studio at The Factory in Florence, Alabama.

100% Organic Cotton Jersey

Scientifically speaking, the cotton plant is a shrub that occurs naturally in tropical and subtropical regions around the world and—according to some—has been cultivated for more than eight thousand years. Its soft, fluffy staple fiber grows in a boll around the cotton seeds and is spun into a yarn that's either woven into the natural cloth most widely used in garments today or knitted into a variety of fabrics like the cotton-jersey fabric that is the basis of our work at Alabama Chanin.

The original varieties of cotton (genus *Gossypium*) came in a range of colors from mocha and tan to gray, brown, black, mahogany, red, pink, blue, green, cream, and white. Over the years, mass cultivation nearly wiped out these original, diverse types of cotton, replacing them with gene-manipulated varieties grown with pesticides harmful to the environment. However, today the traditional and colored varieties are being reintroduced by farmers dedicated to preserving part of cotton's ecology and history. In fact, responding to the environmental devastation caused by many modern processes used in growing and manufacturing cotton, this new crop of farmers is returning to old, sustainable methods that not only protect the environment but also add to the ecological diversity of their regions and to the cotton plant itself. These farmers are true artisans and so integral to our process at Alabama Chanin that it would be remiss not to recognize them as equal to the artisans who construct and embellish our garments.

We work primarily with cream- or natural-colored cotton-jersey yardage that we dye using fiber-reactive dyes. Our cotton-jersey fabric is knitted 64" wide but shrinks slightly through the dyeing process to about 60". Because of the fabric's width, most of the projects in this book require very little yardage, and many of them can be made using remnants and scraps you might already have on hand. For smaller projects, cotton-jersey T-shirts can be purchased inexpensively at resale and thrift stores and deconstructed for their fabric.

We wash and/or dye all of our cotton jersey before cutting into it in order to preshrink it. This ensures that finished projects will not shrink.

Thread, Embroidery Floss & Things That Wind

Below is a brief overview of our favorite thread, floss, and kindred notions. Choosing the best material for the task at hand enhances both the process and the product.

Button Craft Thread
This thread is thicker than ordinary sewing thread, and we use it for most of our hand-sewing needs. Made with a polyester core surrounded by very finely spun cotton yarn, it's one of the strongest threads we've found. Its polished finish helps prevent it from weakening as we constantly pull it through the fabric while sewing, it can be washed repeatedly without breaking or wearing, and its thick strand makes it ideal for embroidered embellishment. Currently, this thread comes in about nine colors.

All-Purpose Thread
Designed for use with sewing machines, this standard, lightweight thread is our choice for hand-basting necklines and armholes before constructing garments to keep these curved edges from stretching during construction and for the ruffling techniques described on page 57. We never use this thin thread for embroidery, embellishment, or constructing a garment: Seams and embellishments sewn with all-purpose thread will break, and beads will fall off after wearing and washing a garment only a few times.

Embroidery Floss
Sold in hundreds of colors, embroidery floss is perfect for small intricate embellishments and where color variation is desired. The floss itself is composed of six loosely twisted strands. You can use the floss as is for embellishment techniques like the backstitched reverse appliqué and Climbing Daisy, or you can pull the floss apart and use just a few strands for lighter, thinner embroidered effects. We normally use four strands for embroidery (by threading two strands through the needle and doubling them over). We never use this floss for sewing seams since it's simply not strong enough for this purpose.

Cotton Yarn and Perle Cotton

When we want a heavier look for embellishment or embroidery than we can get with button craft thread or embroidery floss, we often use either a soft cotton knitting yarn or a mercerized, twisted, glossy cotton yarn called perle cotton. Available in a wide variety of colors from many different manufacturers, perle cotton comes in several different sizes, or weights. We use a 3/2 weight for couching and a 5/2 weight for reverse appliqué and garment construction.

Cotton Tape

We use 5mm (³⁄₁₆")-wide cotton tape for ribbon embroidery since it's very durable, and its weight combines well with our cotton-jersey fabrics. Offered in about twenty-five colors, this loosely woven tape is not sold at all fabric and craft stores but can be found through several online suppliers (see the Resources available on our website, www.alabamachanin.com).

Cotton-Jersey Pulls, or Ropes

When cotton-jersey fabric is cut across—that is, against—the grain line (see page 43), its cut edge will roll to the fabric's right side; and, conversely, when the fabric is cut with the grain line, its cut edge will roll to the wrong side. We use this fabric trait to create small "ropes" of cotton jersey from fabric scraps, which we use for our couching technique and for many tasks around the home and office.

To make pulls, or ropes, regardless of how you plan to use them, first cut your cotton-jersey scraps into strips from ½" to 1" wide (depending on how thick you want the rope to be), cutting the strips with or against the fabric's grain according to which way you want them to curl. After cutting a strip, grab one end with each hand, and pull the strip so that it curls into a rope. That's it!

Additional Notions

Beads, sequins, and fold-over elastic are additional notions that we use on a daily basis at Alabama Chanin.

Beads

We use glass beads that are made in the Czech Republic. Making glass beads is among the oldest of human arts, dating back about three thousand years. We use primarily #2 and #3 bugle beads, chop beads (which look like very short bugle beads), and #7 seed beads. In Chapter 5, you'll find information about the needle and thread we use for beading as well as beading techniques.

Sequins

Most sequins are made from metal or plastic. They are available in a wide range of shapes, sizes and colors; however, at Alabama Chanin, we most often use the flat, round plastic versions in three different sizes: small (4mm), medium (5mm), and large (6mm).

Fold-Over Elastic

We use this fantastic grosgrain binding on the waistbands of all of our pull-on skirts. It is made with a central ridge that is easy to fold over to encase a fabric edge. It is about ⅞" wide (before folding) and thin enough to stitch through.

Toolkits

Below are the tools we use most often as we design and manufacture our collections. We suggest that you build your own toolkits over time as your desires, needs, and skills evolve.

Design Tools

- Research library
- Fabric color cards
- Tracing paper, for sketching and designing repeating patterns
- Mechanical pencil
- Eraser
- Colored pencils and markers

Marking Tools

- 6" transparent, gridded, flexible plastic ruler
- 18"–24" gridded plastic ruler
- Tailor's chalk
- Disappearing-ink fabric pen
- Pattern paper or butcher paper

Stitching Tools

- Sewing needles in assorted sizes
- Beading needles (see page 73)
- Thimbles
- Small rubber finger cap, for gripping needles (sold at office supply stores)
- Needle-nose pliers, for helping pull needles through layers of fabric
- Straight pins

Cutting Tools

- Garment scissors, for cutting fabric for large projects
- 5" knife-edge scissors, for trimming fabric
- 4" embroidery scissors, for small detailed work
- Paper scissors
- Seam ripper
- Rotary cutter
- Cutting mat
- Craft knife

Stenciling Tools

- Medium- or heavyweight transparent film (preferably Mylar), for making stencils
- Acrylic pennant felt, for making stencils
- Spray adhesive, for holding stencil in place while painting it
- Permanent or textile markers, for transferring stencil designs to fabric
- Textile paint, for transferring stencil designs to fabric
- Clean spray bottles with adjustable nozzles, for spraying textile paint onto fabric
- Airbrush gun and air compressor, recommended only if/when you become very serious about spray-painting stencil designs onto fabric and other surfaces (see page 17)
- Butcher paper, for masking off areas that don't require stenciling or for placing between layers of a T-shirt to keep paint from bleeding through stencil onto back of T-shirt

Chapter 2
Stencils + Stenciling

Stenciling is a cornerstone of both our design process and our business model. We use stencils as tools to transfer decorative patterns onto projects like dresses, skirts, and pillows. The stenciled patterns are then used by our artisans as guides for positioning embroidery and beading. Because the stencils so effectively guide the design, our artisans don't need to work in our studio. Rather, they can work independently as individual business owners when and where they want, scheduling their work time as they like.

Over the years, we have worked with more than four hundred different stencil designs. In this book, we present eleven of our favorites, which can be used in an endless variety to produce stunning results.

Alabama Studio Design Stencils

Below is a list of pages on which you'll find the eleven stencils featured in this book. You can photocopy the stencil artwork directly from the book or you can download it from our website (www.alabamachanin. com). Alternatively, you can buy any of these stencils cut and ready to use from our online store.

To create the T-Shirt Top at right, follow the instructions for our basic sleeveless T-Shirt Top on page 48 and apply the Anna's Garden stencil to the fabric. For more information about this garment, see our Index of Design Choices starting on page 164.

Anna's Garden

For the projects in this book, the Anna's Garden stencil artwork was enlarged by 306 percent. This artwork can be photocopied and enlarged, or it can be downloaded full-size from www.alabamachanin.com.

Abbies's Flower

Making a Stencil

I've made stencils from poster board, wax paper, and even brown paper bags, but my favorite stencil materials are clear, medium- to heavy-weight Mylar film and acrylic pennant felt. Clear Mylar film, sold at craft stores, is easy to cut. Because you can see through it, you can trace a pattern directly on it with a permanent marker or use spray adhesive to affix a design printed on paper to it. Because this material bends easily, it needs to be stored with care. Acrylic pennant felt, available from specialty catalogs and online stores, is extremely durable, easy to cut with an X-acto or craft knife, and easily stored for continued use.

Supplies

Stencil image of your choice

Mylar film, pennant felt, or your choice of stencil material slightly larger than desired size of stencil image

Spray adhesive

Craft knife with sharp blade (sharpness is key)

Cutting mat

Computer with printer (in conjunction with scanned or downloaded artwork) or copy machine with enlarging function (in conjunction with artwork on paper)

Artwork

Tools for stencil-transfer (see page 17)

For the projects in this book, the Abbie's Flower stencil artwork was enlarged by 186 percent. This artwork can be photocopied and enlarged, or it can be downloaded full-size from www.alabamachanin.com.

1. Choose and Copy Stencil Design

Choose a stencil image. Photocopy your image if it's printed; if it's on a CD or scanned into your computer, enlarge or reduce it to the size you want, and then print it out. If you want the image to be larger than the letter-sized paper used by most home printers and copiers, take it to a copy shop that has a more versatile printer.

2. Spray Stencil Design with Adhesive and Affix Design to Stencil Material

Working in a well-ventilated area and following the instructions on the spray adhesive's label, lightly spray the back of the paper printout of your stencil design with adhesive to keep the image from shifting as you cut. Then affix the paper to your transparent film or pennant felt, positioning it with a border of about 3" all around.

3. Cut Out Stencil Design

Place the film or felt on a cutting mat with your design facing up. Use the tip of a craft knife to cut out all of the design's elements, leaving just the negative space around these elements—your stencil is a negative image of your original piece of art. Work carefully and slowly to avoid injury.

4. Test Stencil

To test a stencil before transferring the image to your project fabric, lay the stencil on top of a piece of paper or transparent film, and transfer the stencil (see page 17) using textile paint to entirely fill in all the areas that you cut out. The resulting image shows exactly how your stencil will look on your fabric. You've also created a backup image that can be cut into a new stencil if the original gets lost or damaged.

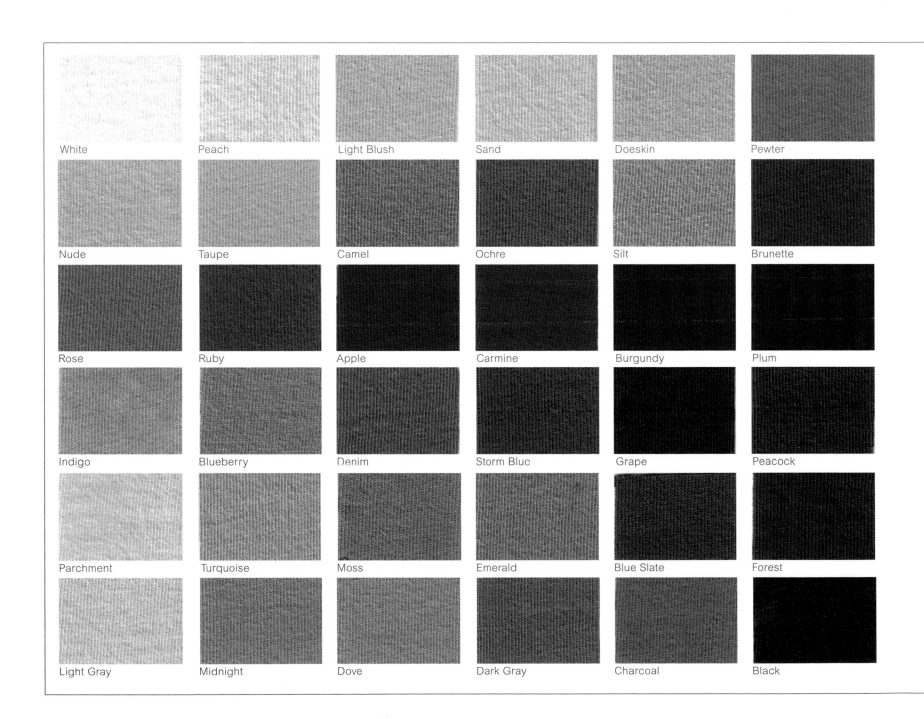

White Peach Light Blush Sand Doeskin Pewter

Nude Taupe Camel Ochre Silt Brunette

Rose Ruby Apple Carmine Burgundy Plum

Indigo Blueberry Denim Storm Blue Grape Peacock

Parchment Turquoise Moss Emerald Blue Slate Forest

Light Gray Midnight Dove Dark Gray Charcoal Black

Fabric and Textile Paint Colors

We currently dye our organic cotton jersey in a wide range of colors. At left are thirty-six of our favorites, including the ones used for the projects and fabrics featured most often in this book.

We use textile paint to transfer stencils to our fabric. While the colors that can be produced by mixing textile paints are limitless, we work primarily with the following fourteen base colors of Createx Airbrush Colors: opaque black, opaque white, opaque blue, opaque sky blue, opaque yellow, opaque red, fluorescent blue, fluorescent hot pink, fluorescent orange, transparent sand, transparent forest green and—when we want to add sheen to a project—pearlized white, pearlized silver, and pearlized red.

By mixing these fourteen colors in a variety of combinations, we create many hues and shades that help us define our patterns. Because our textile paint colors are transparent, each one looks different depending on the color of the fabric to which it is applied. At right, you can see how our cream paint (a mix of transparent sand and opaque white) looks on three different-colored swatches of cotton-jersey fabric.

When mixing paint, always work in small increments, adding only a drop or two at a time since even small drops can change a color dramatically. To make any color more transparent, you can add water. Always test paint colors on a small fabric scrap before you begin to apply them to your project. And always mix more than enough paint for your project (in case you make a mistake or decide to stencil more fabric) because, inevitably, every batch of paint you mix will be slightly different. You can store mixed textile paint, tightly sealed, for about three months. Always test this paint on a fabric scrap to make sure it is working the way you want.

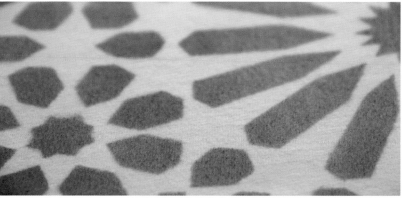

White

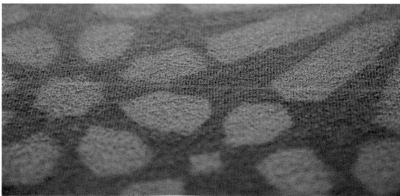

Charcoal

Peacock

Single, Repeating, and Allover Stencil Designs

After choosing a stencil image and creating the stencil itself, you'll need to decide where to place your stencil design on your cut pieces or fabric. For some projects, you might want to transfer a single repeat of your stencil. For example, in the illustrations below, the Fitted Top is shown with a single and double repeat of the Abbie's Flower stencil and with the stencil used as an allover design. If you're working with a relatively small piece of fabric and a large stencil, that single stencil may produce an allover pattern. If you're working with a large piece of fabric and a small stencil, then you'll have to either transfer your stencil design multiple times or make your own large overall stencil. To save time at Alabama Chanin, we like to create 18" x 24" allover stencils because they're large enough to cover the body of most of our tops and yet easy to handle and store. We build a minimum 3" border into the edge of all our stencils to prevent textile paint from "bleeding" beyond the stenciled area and also to strengthen the stencil itself. If you want to create your own allover stencil, make it in the size that works best for you.

Applying Abbie's Flower Stencil to Fitted Top

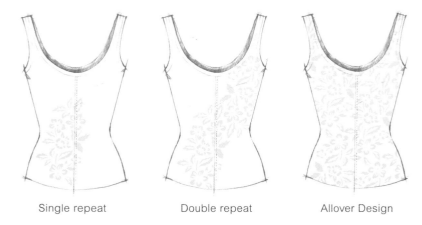

Single repeat   Double repeat   Allover Design

Stencil Transfer

We like to transfer stencil designs to fabric by spray-painting through our stencils. Spray-painting can be done with either a spray bottle or an airbrush gun. Using a spray bottle is a great way to get started transferring designs. If you decide to do multiple projects (and we hope you will), you may want to try using an airbrush gun, which is quicker and can be used repeatedly with different colors and paint mixtures. Some paints require heat to set, or become permanent, while others simply need to be air-dried. Whatever method you choose for applying paint, always test your paint on a scrap of fabric before beginning an actual project, and closely follow all the paint supplier's instructions for safety and/or heat-setting.

If you want to apply paint with a spray bottle, use clean, recycled bottles or new ones with manually adjustable nozzles. Before transferring a stencil to your project fabric, play with various mixtures of paint and water to get the effect that you're looking for (the more water you use, the more transparent the color will be).

If you want to apply paint with an airbrush gun, we recommend choosing the most basic, least expensive, hobby airbrush with a simple air compressor from your local hardware store. For years, we've used a 6-gallon, 150-PSI (pounds per square inch), electric air compressor. Play with the PSI on your compressor to find the perfect setting. This can take some time and concentration to sort out, but once you've adjusted your settings, the airbrush should be easy to use and run smoothly for many years. Each time you finish using an airbrush gun, wash and dry it thoroughly, so it doesn't get clogged with dry paint. Store the paint in airtight containers for up to three months; over time, small particles that can clog your airbrush form in the paint.

Supplies

Stencil

Cotton-jersey fabric

Textile paint

Spray bottle or airbrush gun

Spray adhesive

1. Transfer Stencil Design
After preparing your work surface and stencil and positioning your fabric on a covered work surface, use a spray bottle or airbrush gun to carefully transfer the stencil onto the fabric by spraying paint inside each of the stencil's cut-out shapes. When you've finished spray-painting the entire stencil, carefully remove the stencil and let the paint dry to the touch.

After the paint has dried, if you want to repeat the stencil design or create an allover design on the fabric, reposition the stencil adjacent to the first stenciled motif, spray-paint the stencil again, and repeat this process until you've created the desired design.

2. Dry and Heat-Set If Required
Finish your stenciled design by letting it dry completely. We normally allow 24 hours for our textile paint (which does not require heat-setting) to air-dry before beginning any additional embellishment or sewing. If you're working with paints that require heat-setting to become permanent, always follow the manufacturer's directions in order to prevent the transferred color from washing out. Also note that if you're working with paints that air-dry, we don't recommend washing the stenciled fabric for a minimum of three weeks so that the paint can cure completely (this doesn't apply to heat-set paints).

Varying Your Stencil Effect

On page 17 are instructions for transferring your stencil design to your fabric. Following are some easy variations. Fabrics that are inked with permanent marker can be washed at home; they should not be dry cleaned because the chemicals used will erase the inking. Avoid red markers because they tend to bleed.

The photo at left shows a baby blanket I made for my daughter the year she was born using a simple 48"-x-48" square of cotton jersey. It includes small stanzas about her first days of life. To create your own baby blanket, follow the instructions for Paisley with Poetry & Inking at right.

Inked Stenciling with Ultra-Fine or Fine-Tipped Marker
Use a fine or ultra-fine permanent marker to hand-trace a line around the outside of a stenciled shape to add detail to your project.

Pattern Inking with Fine-Tipped Marker

Use a fine-tipped permanent marker to apply a freehand pattern inside and around any stenciled shape to produce an additional graphic pattern over your stenciling.

Washed-Black Stenciling

Washed-black stenciling softens the appearance of the paint on the fabric's surface. To create this faded effect, dilute three parts black textile paint with one part water. Use a standard spray bottle to transfer stencil to cotton jersey, let dry for 24 hours, and then wash or heat-set following the instructions provided with your textile paint.

Pattern Inking with Ultra-Fine Marker

Use an ultra-fine permanent marker to apply a delicate freehand pattern inside and around any stenciled shape to produce a fine secondary graphic pattern.

Paisley with Poetry & Inking

Follow the instructions for washed-black stenciling above, using the Paisley stencil from page 124. Add inked stenciling with an extra-fine permanent marker, and elaborate with poems or stories.

Chapter 3
Basic Stitches

Featured in this chapter are the basic hand stitches that we use again and again in our designs for construction and embellishment. In Chapter 5 you will find the purely decorative stitches we favor. Of course, there are many more possibilities; if you want to expand your repertoire of stitches, refer to one of the many embroidery-stitch compendiums on the market. Our favorite is *Elegant Stitches* by Judith Baker Montano, which some of our artisans call their "embroidery bible."

Stitch Length and Tension

For most of the projects in this book, you'll sew with a doubled length of thread and you'll make stitches that are between ⅛" and ¼" long. Stitches that are too small will pull through the tiny loops of yarn that make up the cotton-jersey fabric (see "Family of Stitches" on page 28), while stitches that are too big will break more easily or snag as you go about your daily life. Aim not only for uniform stitches and spaces between them but also for even sewing tension. If your sewing tension is too tight, you'll pull the seam too tight as you sew, and it will start to gather up. Conversely, if your tension is too loose, your seam will buckle.

To create the dress at right, follow our instructions for our basic Fitted Dress on page 53 and apply the Spirals stencil to the fabric. For more information about this garment, see our Index of Design Choices on page 164.

Knotting Off

In hand-sewing, your knot anchors and holds your entire seam. Since cotton jersey is a knit fabric made by continuously looping a thin yarn through itself to form a knitted "web," very small holes are formed where the yarn loops. If you knot your sewing thread with a small knot, that knot can pull through any of the small holes in the fabric and might also break the knit fabric's tiny yarn, causing the fabric to "run" and produce an even bigger hole. That's why, in most of our projects, we tell you to double your thread and use a large double knot to anchor the thread (see below).

Another way to ensure the durability of your garment is to leave a ½"-long thread tail after tying off each knot. Wearing and washing your garment will cause these thread tails to shorten over time. For this reason, if you start with long tails, the garment will maintain its original integrity from the first day it was knotted. We like to say we leave long tails so our garments will remain intact for this generation and the next and the next.

One important design decision we make when starting any project is how to handle the knots. There are two options: knots that show on the project's right side (outside) and knots that show on the wrong side (inside). Either of these knots can be used throughout an entire project or combined with the other type (see right).

Tying a Double Knot

After bringing needle up through fabric, make loop with thread, then pull needle through loop, using forefinger or thumb to nudge knot into place, flush with fabric. Then repeat process to make double knot. After making second knot, cut thread, leaving ½" tail.

Alabama Chanin Knots as Decorative Elements

A lovely way to add detail to any stenciled design, garment, or project is to use the Alabama Chanin double knot as an embellishment. In the photos above, you can see our Anna's Garden stencil worked several different ways: without knots showing on the outside (they're all on the reverse), with a few knots on the outside, and with many knots on the outside. The spiral technique (see left and page 83), also called Alabama fur, uses many visible knots to create the effect of fur.

Non-Stretch and Stretch Stitches

We use three categories of stitches in our work: stitches that do not stretch, for construction, reverse appliqué, and other embellishments; stitches that do stretch, for sewing necklines, armholes, and other areas in a project that require "give"; and stitches that are purely decorative, for embellishment, which you will find in Chapter 5.

No matter what stitch we're using, we always "love" our thread to start. I wrote about this process in both *Alabama Stitch Book* and *Alabama Studio* Style and repeat it here because its importance cannot be stressed enough.

When thread is made, microscopic, squiggly cotton fibers are combed in the same direction into two strands, each called a ply. Then one of these plies is twisted to the right (in a so-called "S" twist), while the other strand is twisted to the left (in a "Z" twist). The two plies are finally twisted around each other as tightly as possible, so that, when released, they relax, expand, and together create a single strand of thread. The tension between the two plies explains why thread holds together and also why it sometimes knots as you sew with it. The knotting is caused by excess tension. But there is a way of reducing this tension, a ritual I call loving your thread.

To love your thread, cut a piece twice as long as the distance from your fingers to your elbow. Thread your needle, pulling the thread through the needle until the two ends of the thread are the same length. Hold the doubled thread between your thumb and index finger, and run your fingers along it from the needle to the end of the loose tails. Repeat this several times. What you're doing is working the tension out of the highly twisted thread with rubbing, pressure, and the natural oils on your fingers. In the process, you've also taken a moment to calm the tension in your mind and add just a little bit of love to your project. Now you're ready to tie a knot (see page 21) and start sewing.

Non-stretch stitches shown from top to bottom: straight, or running, stitch; basting stitch; backstitch; and blanket stitch.

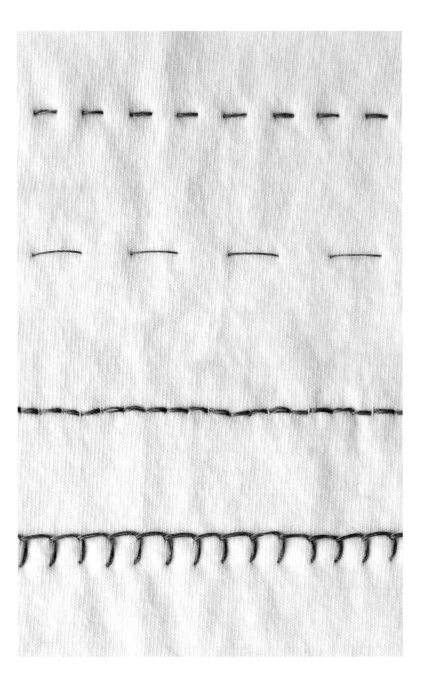

Non-Stretch Stitches

When working with all but one of our non-stretch stitches—the blanket stitch—you'll move the needle in a continuous straight line. In the case of the blanket stitch, you'll use a looping technique to create a straight line.

Straight, or Running, Stitch

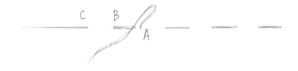

Bring Needle up at A, go back down at B, and come up at C, making stitches and spaces between them same length (about ⅛" to ¼" long).

Backstitch

Bring needle up at A, go back down at B, and come up at C. Then insert needle just ahead of B, and come up at D. Next insert needle just ahead of C, and come up at E. Continue this overall pattern.

Basting Stitch

This stitch is a longer, looser variation of the straight, or running, stitch, and involves simply making both stitches and spaces between them about ½" long.

Blanket Stitch

Bring needle up at A, and hold thread with finger to left of A. Insert needle at B, about ¼" below and to left of A. Come back up at C, directly above B, making sure needle stitches over, not under, thread. Pull up thread, so it lies tightly against thread at C, and repeat process. Continue working stitch, keeping its length and spacing consistent, to complete entire edge or eyelet.

Stitch Sampler

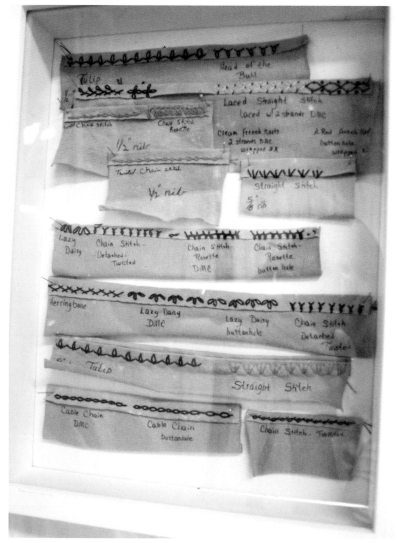

We often create samplers to determine the stretchability of a particular size or length of stitch or to see how various colors or stitches work together. These samplers make a beautiful display when presented in shadow boxes.

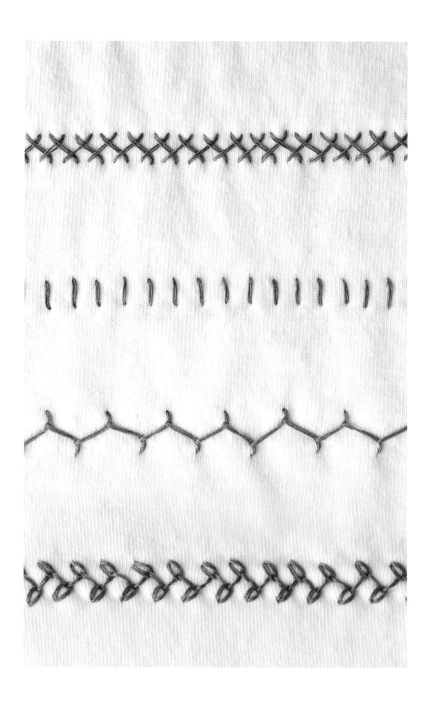

Stretch Stitches

When sewing a stretch stitch, you'll move the needle at an angle, producing stitching lines broken into diagonal or other nonlinear segments, which allow the knit fabric to retain its stretch. All stretch stitches can also be used decoratively and incorporated into embroidery and other fabric embellishment.

Herringbone

Bring needle up at A, go back down at B, come up at C, and go down at D to complete one herringbone stitch. To start next stitch, come back up at new point A, go back down at new point B, and continue working in above pattern to create a row of herringbone stitches, keeping their length and spacing consistent.

Cretan Stitch

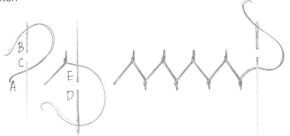

Bring needle up at A, go down at B, and come back up at C, making a downward vertical stitch while bringing needle over thread. Insert needle again at D, and come back up at E, making an upward vertical stitch while bringing needle over thread. Continue to repeat stitch pattern.

Parallel Whipstitch

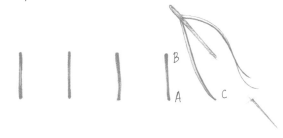

Bring needle up at A, go back down at B, and come up at C, making both stitches and spaces between them ⅜".

Stretch stitches shown from top to bottom at left: herringbone, parallel whipstitch, Cretan stitch, and rosebud stitch.

Rosebud Stitch

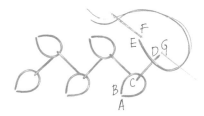

Work this stitch along two imaginary parallel lines, starting by bringing needle up at A. Loop thread above and to right of A, and insert needle back down at B, just to left of A. Bring needle back up at C, pulling thread over, not under, thread loop you made. Go back down at D, and come back up at E. Loop thread below and to right of E, go back down at F, and come up G, pulling thread over, not under, thread loop. Continue repeating stitch pattern, alternating back and forth between parallel lines.

Additional Stretch Stitches

Below are some additional stretch stitches. You should feel free to substitute them in any of the projects in this book.

Chevron Stitch

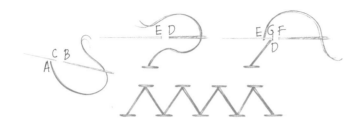

Work stitch along two imaginary parallel lines. Start by bringing needle up at A, go back down at B, and emerge at C. Then insert needle at D, and come back up at E. To form top of chevron stitch, go back down at F, and come up at G. Then move back to imaginary bottom line, and continue working stitch, as before, alternating between top and bottom lines.

Zigzag Chain Stitch

Note that this decorative zigzag chain stitch can also be worked in a straight line as a simple chain stitch.

Bring needle up at A, form thread loop, and go back down at B (very close to A, but not in it). Come back up at C, placing needle's point over thread and pulling thread through. To form next loop, insert needle at D, just inside the first loop to keep it in place, bring needle up at E, placing needle's point over thread, and pull thread through. Continue working this pattern, alternating from side to side.

Snail Trail Stitch

Work stitch along imaginary line, bringing needle up at A, making loop, and going back down at B. Come back up at C, through loop and over thread to secure loop in place.

Featherstitch

This stitch is worked along four imaginary parallel vertical lines. Bring needle up at A, go back down at B, and come up at C, moving needle over thread to make open loop. Repeat process on other side, working downward, and continue to alternate sides until you have completed a row.

Additional stretch stitches shown from top to bottom: chevron stitch; snail trail stitch; zigzag chain stitch; featherstitch; cross-stitch; and double Cretan stitch.

Cross-Stitch

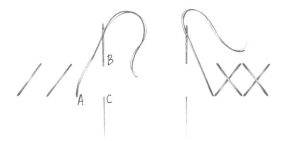

Bring needle up at A, go back down at B, and come up at C; continue this pattern to end of row. Then work same stitch in opposite direction, from lower right to upper left over previous stitches, to form an X.

Double Cretan Stitch

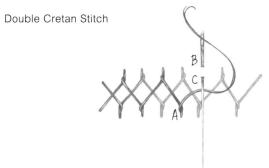

To create the double Cretan stitch, we work the Cretan stitch from page 25 twice across the same line. Simply work your first row of Cretan stitch across your project as required and then work a second Cretan stitch on top of the first row, in a mirror image of your first row to create an interesting diamond shape in the middle. You may also choose to vary the widths of your two Cretan stitches and the color of threads to achieve a variety of stunning results.

Family of Stitches

When looking at cotton-jersey fabric under a microscope, you'll see that a very fine yarn is used to knit rows of stitches that become the basic fabric. However, due to the looping technique involved in producing the knitted fabric, you'll also see that the yarn itself and the negative space around it occupy equal amounts of space. Understanding the "physics" of the fabric at the structural level is key to our work at Alabama Chanin.

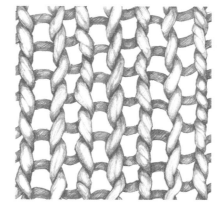

We use button craft thread, the strongest thread we've been able to find, since we want to create garments and other products that will stand the test of time. However, when using a doubled strand of this strong thread to sew together a fabric made of very fine yarn, it's important to understand the physics of their relationship.

I like to use a tale about my Grandfather Perkins to define this concept: As children, my cousins and I spent many summer evenings at my grandfather's house in the country. One by one, as Perk decided each of us was ready, he would induct us into the family with a special ritual. With the rest of us gathered on the screened-in side porch, he would bring the "inductee" out into the yard and ask him or her to pick up as many sticks as possible and carry them to the front porch.

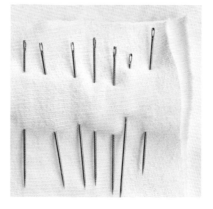

Then Perk would ask the child to take one stick with both hands and break it. When the child easily snapped the twig in two, Perk would say, "That stick is you—breakable."

Then he would tell the child, "Hold as many sticks as you can in your two hands." Once the child had a handful of sticks, Perk would say, "Now break those." Of course, that couldn't be done because of the tensile strength of the bundle. "This," he would say, "is you with your family—unbreakable."

At Alabama Chanin, we similarly incorporate a family of stitches to support the weight of our heavier thread. This is the reason for larger stitches—⅛" to ¼" in length–and also the reason for our double knots. When sewing, always keep in mind that your "family" is there to support the weight of your threads and knots.

Old Wives' Tales and Physics

Over the years, I've heard a lot of old wives' tales around the sewing room, but I've come to learn that many of these tales find truth in everyday life. And as tale after tale has proven true, I've also come to understand that there's reason, or "physics," behind them.

Needle your thread; don't thread your needle.

This makes perfect sense in that the thread is the weaker of the two elements and easily moves or bends. Moving the more stable element—the needle—over the thread to "needle the thread" makes this a simple task.

Long thread, lazy girl.

This is a tale that finds proof on many levels even though it may seem easier or faster to sew with a long thread. First, a longer thread is more prone to knotting as you sew. Second, it seems obvious that if you use an extremely long thread, you'll actually spend more time pulling the thread through your fabric than actually sewing. Third, and most important, as you sew, the thread is being abraded with every stitch you take; and naturally the thread closest to the needle experiences the most abrasion since it is pulled through the fabric more often than the thread closest to the knotted end. Because of this abrasion, the thread closest to the needle becomes weaker and weaker as you sew and sometimes breaks. If you're sewing on a project that might take months to complete, you certainly don't want weakened thread to be apparent after your project's first washing.

Your thread should never be longer than from your fingertips to your elbow.

The best thread length to sew with varies according to individual body size, but it should be about the same as the length from your fingertips to elbow, where the physical action of sewing occurs. I follow this rule but add a couple extra inches to my thread length to accommodate the inch or so needed for my knot and to have "one to grow on."

The end you cut is the end you knot.

This old wives' tale tells us to thread the needle with the end of the thread that comes off the spool and knot the end that you cut from the spool. This is because it's simply easier to thread a needle with the end that comes from the spool.

The physics behind this tale has to do with the "twisting method" involved in manufacturing thread (see page 22). The thread's twist runs in the direction from the loose end towards the spool. If you're unsure of the twist's direction, have a look at a cut length of thread under a magnifying glass, and you'll see that one end is pointed and the other end flares open slightly. Thread the pointed end.

Needle your thread, and then love it good.

This is the one tale that I believe we wrote at Alabama Chanin; and we've found it especially true when sewing with a doubled length of thread. By threading your needle and loving your thread (see page 22), you're training those two lengths of thread to lie side-by-side and begin to behave like "twins." The more the two strands are tamed to lie beside one another, the smoother your sewing path will be since you've created "the path of least resistance."

The knot is the tie that binds.

In hand-sewing, there's no looping action of the thread, as there is on a sewing machine, to help a seam stay forged together. Your sole means of anchoring a hand-sewn stitching line and keeping it in place are the knots at the beginning and end of each seam or embroidery and your correct sewing tension. Just as you created the "path of least resistance" to sew a seam by loving your thread (see above), you've also created a path of least resistance for that seam to come undone if you lose the knot at the beginning or end of your work. For this reason, always double- (and triple-) check that you have the perfect knot, or "the tie that binds."

Chapter 4
Basic Garments + Accessories

In this chapter we provide an overview of the master patterns for four basic styles—the T-Shirt Top/Bolero and the Fitted Dress/Skirt/Tunic—each of which can be cut out and constructed in numerous ways to create a variety of garments. You will also find variations on these garments that can be achieved by attaching one or more ruffle borders in varying lengths. Additionally, you'll find patterns and instructions for making a few of our favorite accessories—the Poncho, Tied Wrap, Bucket Hat, and Fingerless Gloves. Make all of these pieces in their most basic, unembellished form—as shown in this chapter—or mix and match them with the embellishment techniques from Chapters 5 through 8 to create unique designs that fit your individual style. Chapter 9 features our interpretations of the possibilities.

Shown here are three basics: the Bolero—with both long and cap sleeves—the T-Shirt Top with long fluted sleeves, and the Long Skirt.

T-Shirt Top

This is our classic T-Shirt Top, which we also use for a sleeveless shell. The top has a slightly fitted waist and a longer length than a standard T-shirt. Over the years we've offered this T-shirt with various sleeve styles, three of which are provided with the pattern: long fluted, short, and cap. The top measures about 26" from the shoulder and can be shortened or lengthened easily at the bottom edge. A single-layer garment requires about 1 yard of 60"-wide fabric.

Bolero

The Bolero is our favorite cover-up at Alabama Chanin. It works well with the sleeveless fitted T-Shirt Top, the Fitted Dress, or any sleeveless garment of your choice. Included as part of the T-Shirt Top pattern, the Bolero can be constructed with sleeves, if you want, using one of the T-Shirt Top's sleeve variations. The Bolero measures about 11" from the center back to the bottom back edge. A single-layer garment with long sleeves requires 1 yard of fabric. Sleeveless boleros can be cut from ⅜ yard of fabric.

Sleeveless

Cap sleeves

Sleeveless Bolero

Bolero with cap sleeves

Short sleeves

Long fluted sleeves

Bolero with short sleeves

Bolero with long fluted sleeves

Fitted Dress/Skirt/Tunic

This pattern combines one of our favorite sleeveless dresses with a skirt we love. The pattern can be cut at various points to create seven different styles—a Short Fitted Dress, Long Fitted Dress, Fitted Top, Fitted Tunic, Short Skirt, Mid-Length Skirt, and Long Skirt—which can then be layered and worn in various ways.

Long Fitted Dress

This long version of our Fitted Dress is a favorite in our Bridal Collection. It's flattering and comfortable with a clean silhouette and a small train, which can be elaborately embroidered for special occasions. The dress measures about 55" from the shoulder to the front hem and 61" from the shoulder to the back-train hem, and can be shortened or lengthened easily at the bottom edge. A single-layer garment requires about 2 yards of fabric.

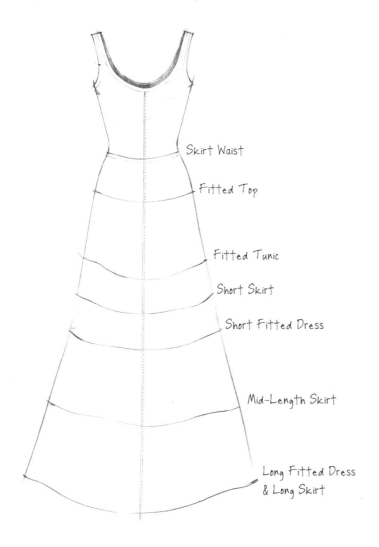

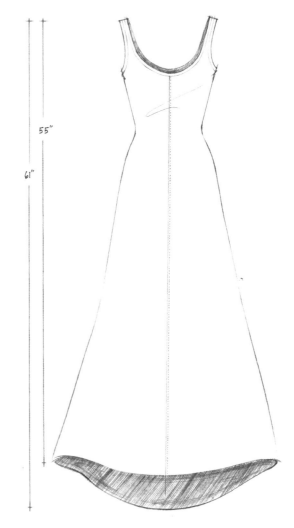

Short Fitted Dress

This dress has thin straps, a scooped neckline, and a shaped center-front seam to give lift in the bust. It's fitted at the waist and flares slightly at the high hip. The dress measures about 40" from the shoulder to the hem and can be shortened or lengthened at the bottom edge. A single-layer garment requires about 1½ yards of fabric.

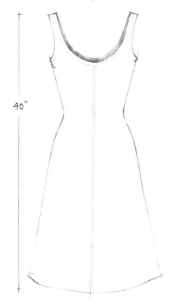

Fitted Top

This short version of our Fitted Dress looks great with jeans and any of our skirt lengths. The top measures about 24" from the shoulder and can be shortened or lengthened easily at the bottom edge. A single-layer garment requires about 1 yard of fabric.

Fitted Tunic

With a length between that of the Fitted Top and the Short Fitted Dress, the Fitted Tunic looks great on its own as well as layered with shorter blouses and jackets. The tunic measures about 30½" from the shoulder to the bottom edge and can be shortened or lengthened easily at this bottom edge. A single-layer garment requires about 1 yard of fabric.

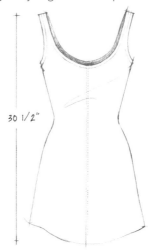

Short Skirt

This simple, pull-on Short Skirt is great for just about every occasion, and its small cut pattern pieces are perfect for elaborate embroidery. The skirt measures about 21" from the waist to the bottom edge and can be shortened or lengthened easily at this bottom edge. A single-layer garment requires about 1 yard of fabric.

Mid-Length Skirt

With a length between the Short Skirt and the Long Skirt, this version falls below the knee and flairs slightly. The skirt measures about 32" from the waist to the bottom edge and can be shortened or lengthened easily at this bottom edge. A single-layer garment requires about 1¼ yards of fabric.

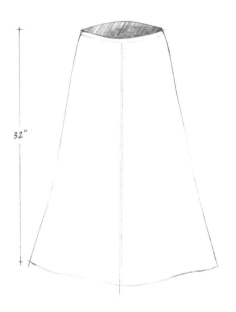

Long Skirt

This Long Skirt with a small train has become a staple in nearly every collection we do at Alabama Chanin. Depending on how it's embellished and accessorized, it can be worn as everything from eveningwear to a beach cover-up. The train on the back of the skirt is 4" longer than the front hem. Measuring about 40½" from the waist to the front hem and 44½" from the waist to the back train, this skirt can be shortened or lengthened easily at the bottom edge. A single-layer garment requires about 1½ yards of fabric.

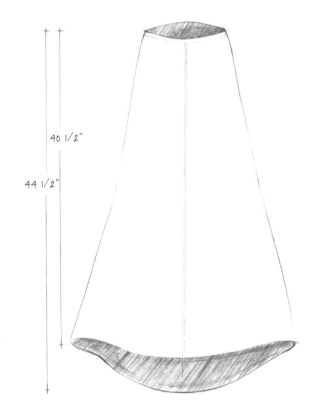

Poncho

This versatile cover-up is just a strategically sewn rectangular piece of fabric. A single-layer garment requires 2 yards of fabric.

Bucket Hat

This versatile piece has a simple form that works beautifully either entirely plain or heavily embellished. Make this hat single-layer for a transitional piece, and add heavy embroidery for winter or eveningwear. This hat requires ½ yard of fabric or can be made with fabric scraps.

Tied Wrap

Adding two ties to a simple rectangle enables it to be transformed when wrapped and worn into a bolero-like garment that's great both summer and winter for beachwear and eveningwear. This piece is perfect for heavy embellishment requiring two or more layers of fabric since these embellishments give the piece weight and structure when worn. A double-layer garment requires 2 yards of fabric.

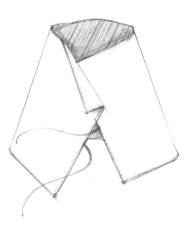

Fingerless Gloves

Make these beautiful gloves when you want to to test out new colorways or embellishments, or when you need a quick gift. The pattern requires only ⅜ yard of fabric or can be made with fabric scraps.

13 ½"

T-Shirt Top/Bolero Master Pattern

The master pattern for the T-Shirt Top at the back of the book includes the Bolero pattern. Both garments can be made sleeveless or with any of the three sleeve variations provided: long fluted, short, and cap sleeves.

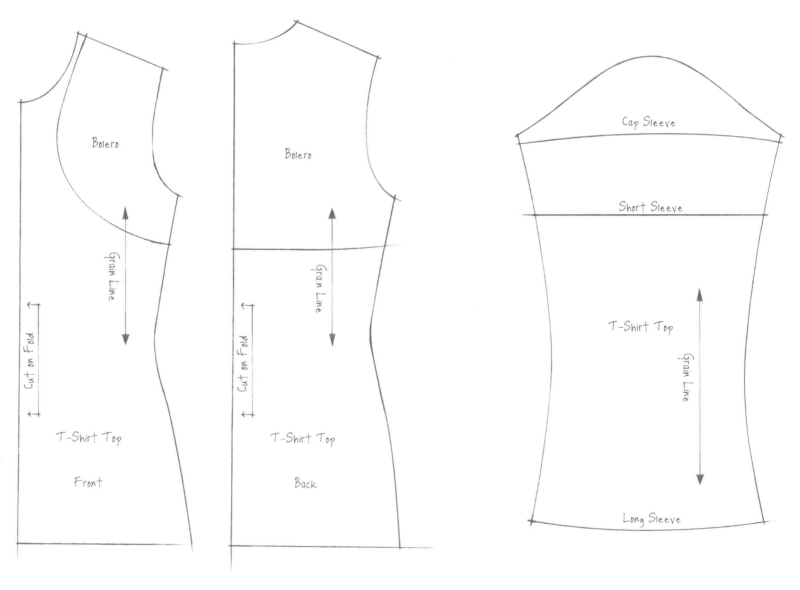

Fitted Dress/Skirt Master Pattern

The master pattern for the Fitted Dress/Skirt at the back of the book can be used to make the Fitted Dress in all its seven variations—Short Fitted Dress, Long Fitted Dress, Fitted Top, Fitted Tunic, Short Skirt, Mid-Length Skirt, and Long Skirt—all of which are featured in Chapter 9.

Poncho Master Pattern

The pattern for the Poncho is a simple 22" x 54" rectangle.

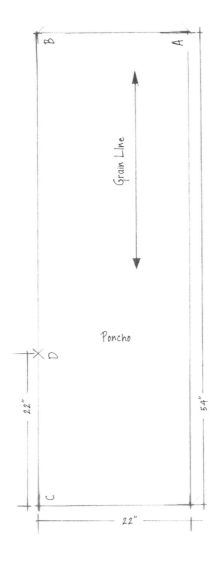

Tied Wrap Master Pattern

The Tied Wrap pattern is a 21" x 48" rectangle with markings for two 18"-long ties to be attached after construction.

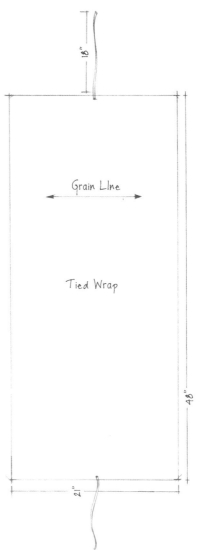

Bucket Hat Master Pattern

Enlarge the pattern pieces by 225% to create the Bucket Hat for an adult and by 176% for a child's version.

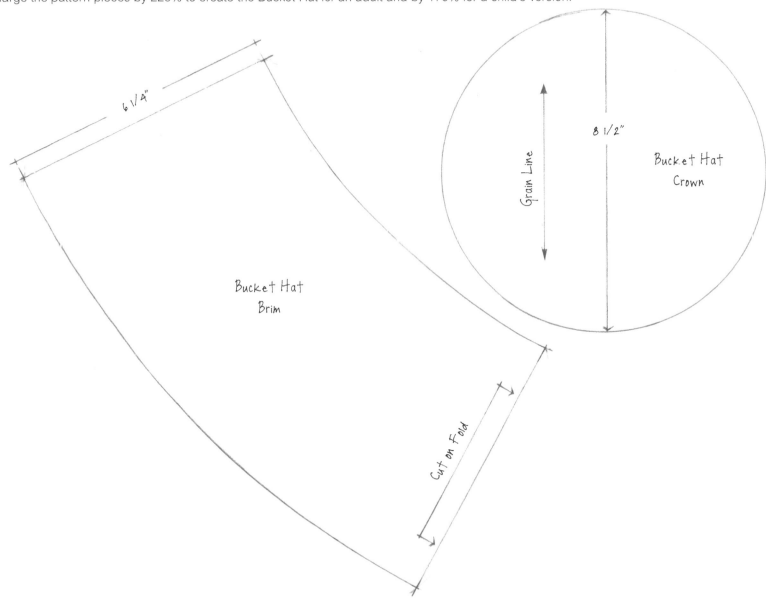

6 1/4"

Bucket Hat
Brim

Cut on Fold

Grain Line

8 1/2"

Bucket Hat
Crown

Fingerless Gloves Master Pattern

Enlarge the pattern below by 317% to fit an average woman's hand.

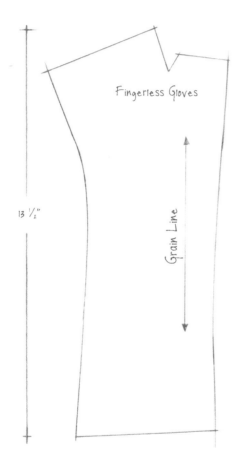

Fingerless Gloves

Grain Line

13 ½"

Prepare for Cutting

Before cutting out a project, you need to decide if you are making it single- or double-layer and what size you want to make. Below are the guidelines we follow when making these decisions.

Single- or Double-Layer?

Single-layer garments are lighter in weight and usually make more sense for warm weather. Double-layer garments are warmer and offer more support, especially at the bust. Some techniques, such as appliqué and beading, can be worked on either single- or double-layer garments however, a garment embellished with extremely heavy beading will have better support if made double-layer. Other techniques, like reverse appliqué, require working on a double layer of fabric (though sometimes we trim away part of the backing layer to reduce the fabric's weight). Obviously, a double-layer garment requires twice the amount of fabric as a single-layer garment. When cutting out a pattern, you'll work either with the fabric opened out flat as a single layer or with it folded double-layer. The directions for each project in this book tell you which way to lay out the fabric for cutting.

Finding Your Size

When it comes to picking a pattern size, everyone has a personal preference. At Alabama Chanin, we recommend a slightly snug fit to start because, over the course of several wearings and washings, a cotton-jersey garment relaxes and begins to take on your body's shape. We make an exception for projects that are heavily embellished with decorative stitching and beading since the embellishment tends to limit the cotton jersey's ability to stretch. In this case, we suggest choosing one pattern size larger than your regular size. Whatever your fit preference, look at the size chart below to see our measurements for each pattern size (note that only chest and waist measurements are given since the base pattern's flared dress and skirt accommodate a wide range of hip sizes). One way to ensure that your garment will fit you perfectly is to make a basic, unembellished version first, try it on for size, and then adapt your next garment accordingly.

	XS	S	M	L	XL
Size	0–2	4–6	6–8	10–12	14–16
Chest	28–30	30–32	32–34	36–38	40–42
Waist	23–24	25–26	27–28	30–32	33–35

Approximate Yardage Required for Basic Garments & Accessories

Note: This chart refers to 60"-wide cotton-jersey fabric and the yardage needed. For double-layer garments, we have provided the total number of yards you will need. If you want to use two different colors for your outer and backing layers, follow the yardages for single-layer garments for each color.

Garments and Accessories	Single-Layer	Double-Layer
T-Shirt Top – Sleeveless, Cap, or Short Sleeves	1 yard	2 yards
T-Shirt Top – Long Sleeves	2 yards	4 yards
Bolero – Sleeveless, Cap, or Short Sleeves	⅜ yard	¾ yard
Bolero – Long Sleeves	1 yard	2 yards
Short Fitted Dress	1½ yards	3 yards
Long Fitted Dress	2 yards	4 yards
Fitted Top	1 yard or less	2 yards or less
Fitted Tunic	1 yard	2 yards
Short Skirt	1 yard or less	2 yards or less
Mid-Length Skirt	1 yard	2 yards
Long Skirt	1½ yards	3 yards
Poncho	2 yards	4 yards
Tied Wrap	1½ yards	3 yards
Bucket Hat	½ yard	1 yard
Fingerless Gloves	⅜ yard	⅜ yard

Cutting Out a Pattern

Cutting out the pattern is an important step that requires precision since it will be the foundation of your project. Below are instructions for cutting out a pattern using cotton-jersey fabric (see page 7).

Supplies

Project pattern

Cotton-jersey fabric

Tailor's chalk or disappearing-ink fabric pen

Paper scissors

Garment scissors

1. Choose Size and Cut Out Pattern

The garment patterns at the back of this book provide five sizes (from XS to XL) in which the garment can be made. Decide which size garment you want to make (see page 41), photocopy or trace the pattern, and use your paper scissors to cut out the photocopied or traced pattern in your desired size.

2. Prepare Fabric for Cutting and Stitching

It's important to prevent the cotton-jersey fabric from stretching as you cut and work with it. To do this, when laying out the fabric on your work surface, don't stretch it or smooth it out by pulling on it; instead lightly pat the cotton jersey into place with your fingertips. The directions for each project will tell you whether to lay out and cut your fabric single- or double-layer (see page 40).

3. Transfer Pattern to Fabric

Lay your paper pattern pieces on top of your fabric, making sure the pattern's marked grain line runs in the same direction as the fabric's grain line (see page 43). This is important because, for example, when cutting out a camisole, you want the grain line on the cut fabric pieces to run vertically from the neckline to waistline, so the fabric can stretch around your body.

As you trace around your pattern piece with tailor's chalk or a disappearing-ink fabric pen, hold the pattern in place with the palm of your hand (or with pattern weights or even canned goods). We prefer holding or weighting the pattern to pinning it on the fabric, which, in the case of cotton jersey, often skews the fabric and makes the cutting uneven. We've also found that this holding/weighting strategy prevents nicking and tearing the pattern, which often results from pinning it in place.

Since many of the projects in this book have more than one pattern piece, we suggest cutting out all the pieces at once. Laying out all the pattern pieces together on your fabric before you start cutting enables you to figure out how to place the pieces to use your fabric most efficiently.

4. Cut Pattern from Fabric

Using garment scissors, cut out the pattern pieces as called for in your project directions, trying your best to cut just inside the chalked or penned line you traced around the pattern. By cutting away all of the visible chalk or ink (but not cutting beyond the marked line), you'll help ensure a perfect fit.

Paper patterns are among the most valuable assets in the Alabama Chanin studio. Over the last decade we have developed more than five hundred different styles that are the basis of our couture collections.

Finding the Grain Line

In a knit fabric, the term *grain line* refers to the direction of the stitches making up the fabric, which, in the case of cotton jersey, typically run vertically along the fabric's length. If you look closely at cotton jersey's right side, you'll see straight vertical columns of stitches that make up the grain line. On cotton jersey's wrong side, you'll see a series of little loops. To cut the fabric with the grain, align your scissors or rotary cutter with the fabric's grain, that is, with its vertical columns of stitches. Working with the correct grain line enhances a finished garment's stretchability at key points (for example, at the armholes, across the bust, and at other parts of the body where we need a bit of stretch).

Right Side

Wrong Side

Seams

For each garment we create at Alabama Chanin, we choose whether to expose or hide the seam allowances and whether to make what we call "floating" or "felled" seams. When the seam allowances are exposed on the garment's right side, or outside, they tend to highlight the garment's structure; and, conversely, when the seam allowances are hidden inside the garment on the wrong side, the look of the garment is more streamlined.

Floating seams are seams whose allowances are not sewn down and, hence, stand a little away from both the seam line itself and the base fabric, giving the garment a rough, deconstructed look. Felled seams, whose seam allowances are sewn down, lie flat and are stronger than floating seams, thanks to the extra line of stitches that felling requires; they also produce a more structured, refined look than floating seams. As a general rule, regardless of whether we're sewing exposed or hidden seams, we stitch the seam line ¼" from the fabric's cut raw edge. If a different seam allowance is called for in a given project, the directions will tell you.

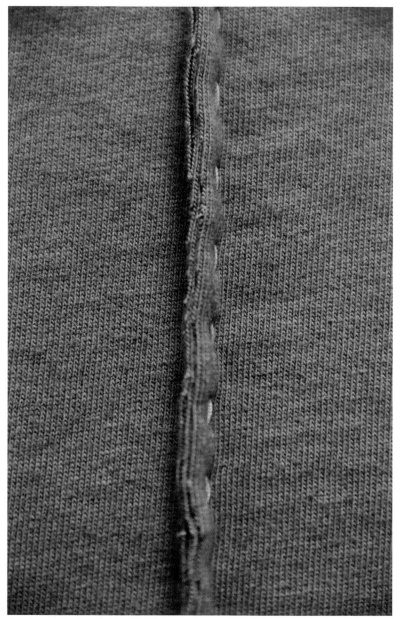

Floating seam on right side

Floating Seams

If you're using floating seams in a project, it's important to make sure that every seam floats, even when one seam intersects another seam. To stitch a floating seam that intersects another floating seam, stitch under the sewn seam being intersected so that its seam allowances continue to float, and then begin stitching your new seam as usual on the other side of the intersected seam.

Floating Seam on Right Side

To make a floating seam that shows on the right side of a project, pin the two cut fabric pieces being seamed with their wrong sides together; then stitch the seam on the fabric's right side. The resulting seam will be visible on the project's right side, or outside.

Floating Seam on Wrong Side

To make a floating seam that's hidden on the inside, or wrong side, of a garment or project, pin the two cut fabric pieces being seamed with their right sides together; then stitch the seam on the fabric's wrong side. The resulting seam will be only visible on the project's wrong side.

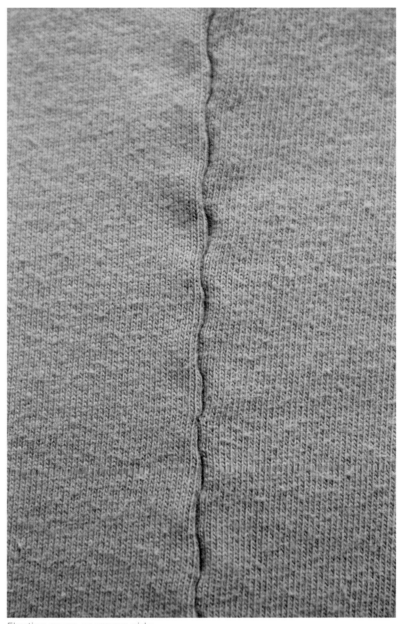

Floating seam on wrong side

Felled Seams

To decide which direction to fold and fell your seams, start at the center front or center back of your garment, and always fell seams toward the side of your body. As a general rule, side seams should be felled toward your back.

Felled Seam on Right Side

To sew a felled seam on the right side of a project, start by sewing a floating seam on the right side (see page 45). Then fold the finished seam's allowances over together to one side, and stitch them down on the right side of the project with a parallel row of stitches ⅛" from the allowances' cut edges or down the center of the seam allowances. The resulting seam will be visible on the project's right side.

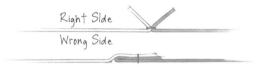

Felled Seam on Wrong Side

To sew a felled seam on the wrong side of a garment or project, start by sewing a floating seam on the wrong side (see page 45). Then fold the finished seam's allowances over together to one side, and stitch them down on the wrong side of the project with a parallel row of stitches ⅛" from the allowances' cut edges or down the center of the seam allowances. The resulting seam will be visible on the project's wrong side.

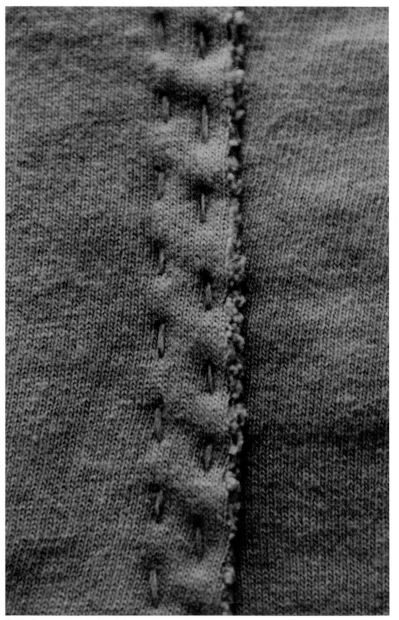

Felled seam on right side

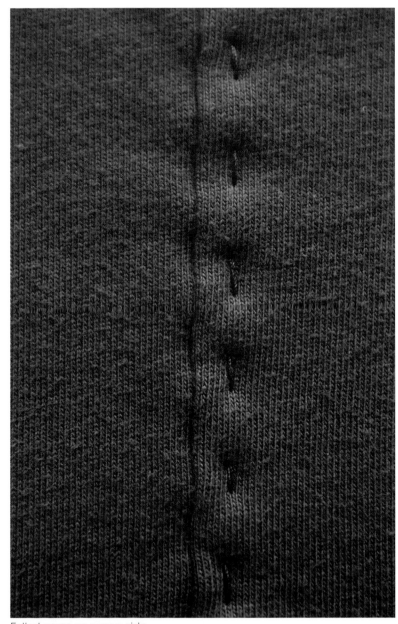

Felled seam on wrong side

Wrap-Stitching Seams

To anchor your seam and ensure that it stays flat with no hint of gathering or pulling, begin and end it by wrap-stitching the fabric's raw edges. To do this, wrap a loop of thread around the edge of the fabric at the first and last stitches, as shown in the drawing below.

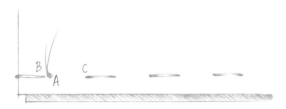

To start a seam, make double knot (see page 21), and insert needle at point A. Wrap thread around side of fabric to front, insert needle back in at B (right next to A), then come up at C, and stitch seam. Also wrap-stitch end of seam before knotting off.

Instructions for Constructing Garments

Before adding a new garment style to a collection, we test the pattern for fit and appeal in its most basic version—without embroidery or embellishment. Over the years, we've come to call these unembellished garments our "basics," and they've become the basis for my personal wardrobe. Below you'll find the instructions for making basic unembellished garments using felled seams on the wrong side (see page 46). Note that the construction process for these garments is the same whether you choose to make them single- or double-layer. Full instructions for making a single-layer T-Shirt Top and a double-layer Bolero are provided at the outset, with instructions for the remaining patterns referring back to these basic patterns for the steps that are the same.

T-Shirt Top

The instructions below are for a single-layer T-Shirt Top. At right it is shown both sleeveless and with long, fluted sleeves.

Supplies

T-Shirt Top pattern (see pattern sheet at back of book)

1 yard of 60"-wide cotton-jersey fabric (2 yards for long-sleeved version)

Paper scissors

Garment scissors

Rotary cutter and cutting mat

18" transparent plastic ruler

Tailor's chalk or disappearing-ink fabric pen

Hand-sewing needle

Button craft thread

All-purpose sewing thread

Pins

1. Prepare and Cut Out Pattern Pieces

Photocopy T-Shirt Top pattern, and cut photocopied pattern to desired size, cutting as close as possible to black cutting line. The T-Shirt Top pattern has two main pattern pieces and an optional sleeve in three styles, with a ¼" seam allowance built into all pattern edges. Place T-Shirt Top pattern front on top of your folded yardage, making sure pattern and fabric grain lines (see page 43) run in same direction. With tailor's chalk, trace around pattern's edges, remove pattern, and cut out traced pattern piece, cutting just inside chalked line to remove it entirely. Repeat this step with remaining cotton jersey with your T-Shirt pattern back. You'll now have a total of two cut pieces for your T-Shirt. If you want sleeves, repeat process above with double layer of fabric and T-Shirt sleeve pattern of your choice to cut out two sleeves.

2. Baste Neckline and Armholes

To ensure that neckline and armholes don't stretch while you're constructing the T-Shirt, use a single strand of all-purpose thread to baste around the neckline and armhole edges of each cut piece.

3. Add Stenciling and/or Embroidery

If your project calls for stenciling (see page 10), add this to the right side of cut T-Shirt pieces, and let stenciled images dry thoroughly before proceeding. Add embellishment as desired. If you're adding beading, avoid beading in ¼" seam allowance.

4. Prepare for Construction

Pin front and back of T-Shirt together at shoulder, with right sides together. (Or, if you want to make seams visible on garment's right side, position cut pieces with wrong sides together.)

5. Sew Shoulder Seams

Thread your needle, love your thread, and knot off (see page 21–22). Using a straight stitch, sew pinned pieces together at shoulder, starting at top edge of T-Shirt's armhole and stitching ¼" from fabric's cut edges across to neckline. Begin and end seam by wrap-stitching (see page 47) its edges to secure them. "Fell" (see page 46) your seam towards back of your T-Shirt and topstitch allowances ¼" from cut edges (down center of seam allowances), using a straight stitch and wrap-stitching the beginning and end of the seam.

6. Add Sleeves (optional)

Pin cut sleeves to T-Shirt armholes with right sides together and matching sleeve's front armhole edge with front of T-Shirt and sleeve's back armhole edge with back of T-Shirt. Pin pieces together securely, working in excess fabric with pins. Thread a needle, love your thread, and knot off. Using a straight stitch, stitch pinned pieces together at armhole, wrap-stitching both ends of seam. Either leave seam floating, or fold seam allowances toward sleeve and fell them down center.

7. Sew T-Shirt Body at Side Seams

Turn T-Shirt wrong side out and pin together front, back, and optional sleeves at side seams. If making a sleeveless garment, join side seams with a straight stitch, wrap-stitching seams. For a garment with sleeves, similarly use a straight stitch and wrap-stitch seams. Start stitching at the bottom edge of T-Shirt's hem and sew side and sleeve seams in one continuous pass, which produces a better-fitting armhole than sewing side seams and sleeves separately and then inserting sleeves into armholes. After stitching side/sleeve seam, fold seam allowances toward back, and fell the seam.

8. Bind Neckline and Armholes, If Sleeveless

Use rotary cutter, cutting mat, and large plastic ruler to cut 1¼"-wide strips of leftover fabric across grain to use for binding neckline and armholes. You'll need a total of about 60" of cut strips. Attaching binding will be easier if you cut one continuous piece—for example, cut one neckline binding strip that's long enough to go around entire neckline.

Use an iron to press each cut binding strip in half lengthwise, with wrong sides together, being careful not to stretch fabric as you press it. Starting at T-Shirt's center-back neckline, encase neckline's raw edge inside folded binding, pinning or basting binding in place as you work (binding strip's raw edges will show). If binding strip isn't long enough to fully cover edge you're binding, add new binding strip by overlapping strips' raw edges by ½". At center-back point, similarly overlap binding's raw edges by ½" to finish, trimming away excess binding.

Using stretch stitch of your choice (see page 25 and 26), sew through all layers and down middle of binding. Repeat process to finish each armhole on a sleeveless garment. Remove or simply break neckline and armhole basting stitches by pulling gently on one end of thread. If some basting stitches are embedded in binding, it's fine to leave them.

Bolero

Our Bolero is shown in the photo on page 30 with both cap and long sleeves. The instructions below can be used for either a single- or double-layer garment.

Supplies

Bolero pattern (combined with T-Shirt Top pattern at back of book)

Long-Sleeve Bolero: 1 yard each of 60"-wide cotton-jersey fabric in two colors, for top and backing layers

Sleeveless, Cap-or Short-Sleeve Bolero: ⅜ yard each of 60"-wide cotton-jersey fabric (or fabric scraps) in two colors, for top and backing layers

Paper scissors

Garment scissors

Rotary cutter and cutting mat

18" transparent plastic ruler

Tailor's chalk or disappearing-ink fabric pen

Hand-sewing needle

Button craft thread

All-purpose sewing thread

Pins

1. Prepare and Cut Pattern
Photocopy the Bolero pattern, and cut the photocopied pattern to your desired size, cutting as close as possible to the black cutting line. The Bolero pattern has two main pattern pieces and an optional sleeve in three styles, with a ¼" seam allowance built into all the pattern edges.

2. Cut Top-Layer Pattern Pieces
Lay out your top-layer fabric flat and fold to create two layers. Then place the Bolero pattern front on top of the folded yardage, making sure the pattern and fabric grain lines run in the same direction. Use tailor's chalk to trace around the pattern's edges, remove the pattern, and use the garment scissors to cut out the traced pattern, cutting just inside the chalked line to remove it entirely. You'll now have your two front pieces. Repeat this step on the remaining yardage with your Bolero pattern back, which will give you one back piece, or a total of three pieces for the garment's top layer.

If you want to add sleeves to your Bolero, cut out the sleeve pattern you want from the remaining double-layered fabric to get two sleeves.

3. Cut Backing-Layer Pattern Pieces (for double-layer garment)
Lay out the 1 yard of fabric for your backing layer flat, and fold it to create two layers. Then repeat the process in Step 2 to cut three backing-layer pieces for the Bolero body.

If you're adding sleeves to your Bolero, cut two backing-layer sleeves from the remaining double-layer fabric.

4. Baste Bolero Body and Armholes
To ensure that the Bolero's curved shape and armholes on your cut fabric pieces don't stretch while you're constructing the garment, use a single strand of all-purpose thread to baste all the way around the neckline, body, and armhole edges of each cut piece, as noted on the pattern piece.

5. Add Stenciling (optional)
If your project calls for stenciling, add the desired stenciling on the right side of the Bolero's top layer only, and let the image dry thoroughly.

6. Pin Pattern's Top and Backing Layers (for double-layer garment)
Align each cut top-layer piece on the corresponding backing-layer

piece, with both fabrics facing right side up, and pat the layers into place with your fingertips so that their edges match. Securely pin together the two layers of each piece.

7. Add Embellishment (optional)
Complete all the embellishments that you want to add to your project. If you're adding beading, be sure to avoid beading in the ¼" seam allowance.

8. Prepare for Construction
Once you've completed your embellishment, begin constructing the Bolero by pinning the front and back panels together at the shoulders, with right sides together.

9. Sew Shoulder Seams
Thread your needle, love your thread, and knot off (see pages 21–22). Using a straight stitch, stitch the pinned pieces together at the shoulder, starting at the top edge of the Bolero's armhole, stitching ¼" from the fabric's cut edges across to the neckline, and beginning and ending the seam by wrap-stitching it. Fell your seam by folding over the seam allowances towards the back of your Bolero and topstitching the seam allowances ⅛" from the cut edges (down the center of the seam allowances), using a straight stitch and again wrap-stitching the seam.

10. Add Sleeves (optional)
Pin the cut sleeves to the Bolero's armholes with right sides together, matching each sleeve's front armhole edge to the Bolero's front and the sleeve's back armhole edge to the Bolero's back. Pin the edges securely, working in any excess fabric with your pins. Thread your needle, love your thread, and knot off. Using a straight stitch, sew the pinned pieces together at the armhole, wrap-stitching the seam. Note that you can either leave the seam floating or fold the seam allowances toward the sleeve and fell them down the center.

11. Sew Bolero Body at Side Seams
Turn your Bolero wrong side out, and pin together the front, back,

and optional sleeves at the side seams. If you're making a sleeveless garment, join the side seams with a straight stitch, wrap-stitching the seam. If you're making a Bolero with sleeves, similarly use a straight stitch and wrap-stitch the seam, starting your stitching at the bottom edge of the Bolero's hem and sewing together the side and sleeve seams in one continuous pass. Attaching the side and sleeve seams this way produces a better-fitting armhole than sewing the side seams and sleeves separately and then inserting the sleeves into the armholes. For this reason, always follow the sequence in steps 8 and 9 when constructing any cotton-jersey garment with sleeves. After stitching the side/sleeve seam, fold the seam allowances toward the back and fell the seam.

12. Bind Neckline and Armholes, If Sleeveless
Use the rotary cutter, cutting mat, and large plastic ruler to cut 1¼"-wide strips of leftover fabric across the grain to use for binding the neckline and armholes. You'll need a total of about 60" of cut strips for the binding. Note that attaching the binding will be easier if you cut the binding for a particular area long enough to fit that entire area without having to piece the binding—for example, cut one piece of neckline binding that's long enough to go entirely around the entire neckline.

Use your iron to press each cut binding strip in half lengthwise, with wrong sides together, being careful not to stretch the fabric as you press it. Starting at the Bolero's center-back neckline, encase the garment's entire raw edge (the neckline, front edges, and bottom edge) inside the folded binding, pinning or basting the binding in place as you work (note that the binding strip's raw edges will show). If your binding strip isn't long enough to fully cover the edge you're binding, add a new binding strip by overlapping the strips' short raw edges by about ½". When you reach the center-back point again, similarly overlap the binding's short raw edges by about ½" to finish, trimming away any excess binding.

Using the stretch stitch of your choice, sew through all layers down the middle of the binding. Repeat the process to finish each armhole if you've chosen to make a sleeveless version. Remove or simply break the neckline and armhole basting stitches by pulling gently on one end of the thread. If some of the basting stitches are embedded in the binding, it's fine to leave them in place since the thread is broken, and the remaining stitches will not restrict the fabric's stretch.

Short/Long Fitted Dress, Fitted Top, and Fitted Tunic

The instructions below are for the Short Fitted Dress but also apply to the Long Fitted Dress, Fitted Top, and Fitted Tunic (with yardage requirement adjusted accordingly; see pages 32 and 33).

The instructions below can be used for either a single- or a double-layer garment.

Supplies

Fitted Dress pattern (see pattern sheet at back of book)

1½ yards of 60"-wide cotton-jersey fabric in one color, for top layer

1½ yards of 60"-wide cotton-jersey fabric in second color, for backing layer (optional)

Paper scissors

Garment scissors

Rotary cutter and cutting mat

18" transparent plastic ruler

Tailor's chalk or disappearing-ink fabric pen

Hand-sewing needle

Button craft thread

All-purpose sewing thread

Pins

1. Prepare and Cut Pattern

Photocopy the Short Fitted Dress pattern, and cut the photocopied pattern to your desired size, cutting as close as possible to the black cutting line. The pattern has two pieces—a front panel and a back panel—with a ¼" seam allowance built into all the pattern edges.

2. Cut Top-Layer Pattern Pieces

Repeat Step 2 of the instructions for making the Bolero on page 50 to get a total of four pieces for the dress's top layer (note that the optional sleeves mentioned in the Bolero directions don't apply to this dress).

3. Cut Backing-Layer Pattern Pieces (for double-layer garment)

Lay out flat and double-layer your backing-layer fabric, and repeat the process in Step 2 to cut four backing-layer pieces.

4. Baste Neckline and Armholes

To ensure that the neckline and armholes on your cut fabric pieces don't stretch while you're constructing the dress, use a single strand of all purpose thread to baste the neckline and armhole edges of each cut piece, as noted on the pattern pieces.

5. Add Stenciling (optional)

Repeat Step 5 of the Bolero instructions.

6. Pin Pattern's Top and Backing Layers (for double-layer garment)

Repeat Step 6 of the Bolero instructions.

7. Add Embellishment (optional)

Repeat Step 7 of the Bolero instructions.

8. Prepare for Construction

Once you've completed your embellishment, begin constructing the dress by pinning the front panels together at the center front, with the right sides facing. Repeat this process for the dress's two back panels, pinning them together at the center back.

Shown at left is our single-layer Long Fitted Dress. At right is our single-layer Short Fitted Dress.

9. Construct Dress

Thread your needle, love your thread, and knot off (see pages 21–22). Using a straight stitch, stitch the pinned pieces together, starting at the top edge of the center front and stitching ¼" from the fabric's cut edges down to the bottom edge. Be sure to begin and end the seam by wrap-stitching its edges to secure them. Fell each seam by folding over the seam allowances to one side and topstitching them ⅛" from the cut edges (down the center of the seam allowances), using a straight stitch and wrap-stitching the seam. Repeat this process for the center back.

Next pin the shoulder seams with right sides together, and sew the shoulder seams, following the instructions above for sewing the front and back panels. Note that you can either leave the shoulder seams floating or fold the allowances toward the back and fell them down the center. Pin the constructed front and back panels together at the side seams, with right sides together, and follow the instructions above for sewing them together.

10. Bind Neckline and Armholes

Use the rotary cutter, cutting mat, and large plastic ruler to cut 1¼"-wide strips of leftover fabric across the grain to use for binding the neckline and armholes. You'll need a total of about 80" of cut strips for the binding. Use your iron to press each binding strip in half lengthwise, with wrong sides together, being careful not to stretch the fabric while pressing it.

Starting at the dress's center-back neckline, encase the neckline's raw edge inside the folded binding (note that the binding's raw edges will show), pinning the binding in place as you work. Add a new binding strip, as needed, by overlapping the strips' short raw edges by about ½"; when you reach the center-back point again, overlap the binding's short raw edges by about ½" to finish the binding, trimming away any excess binding. Using the stretch stitch of your choice, sew through all layers down the middle of the binding.

Repeat the process above to bind and finish each armhole. Remove or simply break the neckline and armhole basting stitches by pulling gently on one end of the thread. If some of the basting stitches are embedded in the binding, leave them in place since the thread is broken, and the remaining stitches will not restrict the fabric's stretch.

Shown at left is our single-layer Fitted Tunic. At right is the neckline and armhole detail. At far right is our single-layer Fitted Top.

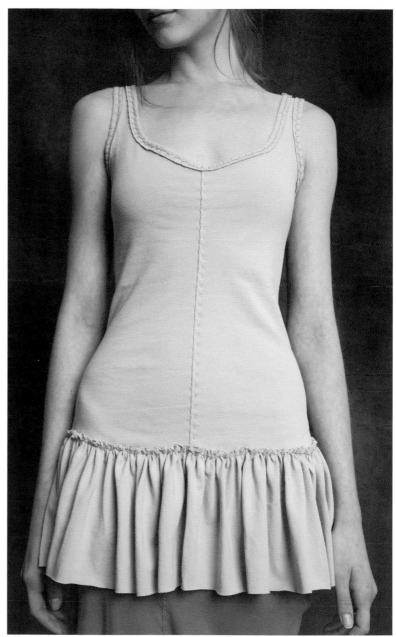

Ruffled Tops and Dresses

Ruffle borders are a great way to add variety to any of the basic garments in this chapter. By appliquéing one or more ruffle borders to a garment's hem and varying the length of the basic garment and/or the ruffles, you can create a multitude of styles and designs.

To create a ruffle border, cut a cotton-jersey rectangle across the grain, making it the width you want and approximately three times the length of the garment's front edge you'll attach it to. Repeat this process for the garment back. Sew the two rectangles together following the seam construction for your garment, creating two side seams. Using all-purpose thread, sew a basting stitch across the rectangle's long top edge, and pull on the ends of the basting thread to gather up the fabric.

Pin the gathered ruffle to the hem of your finished garment, overlapping the garment's bottom edge with the ruffle's top edge by ½", aligning the side seams, and matching the ruffle's length to the garment's width. Distribute and pin the gathers around the hem. Once you have pinned your ruffle into place, use a basting stitch to hold the ruffle. Then attach it permanently with the stretch or decorative stitch of your choice (you can pull out the basting stitches if you want, or leave them hidden in the ruffling).

To create the single-layer Ruffled Tunic at left, we reduced the length of the Fitted Tunic (see page 33) to 21" at center back by shortening it at the bottom edge. After constructing the tunic, we appliquéd a 7"-wide ruffle border to the hem.

To create the Baby Doll Top in the photo at far left, we reduced the length of the Fitted Tunic to 5½" at center back by shortening it at the bottom edge. After constructing the top, we appliquéd a 15"-wide ruffle border to the hem. On page 58 are two Baby Doll Dresses.

Single Ruffle Border

1. Sew basting stitch across ruffle fabric's long, top edge.

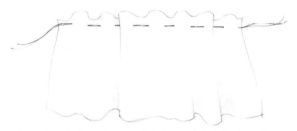

2. Pull on thread ends to gather up fabric, and distribute gathers evenly.

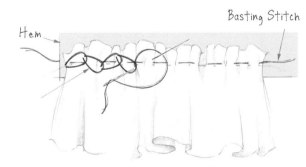

3. Use stretch or decorative stitch to attach ruffle.

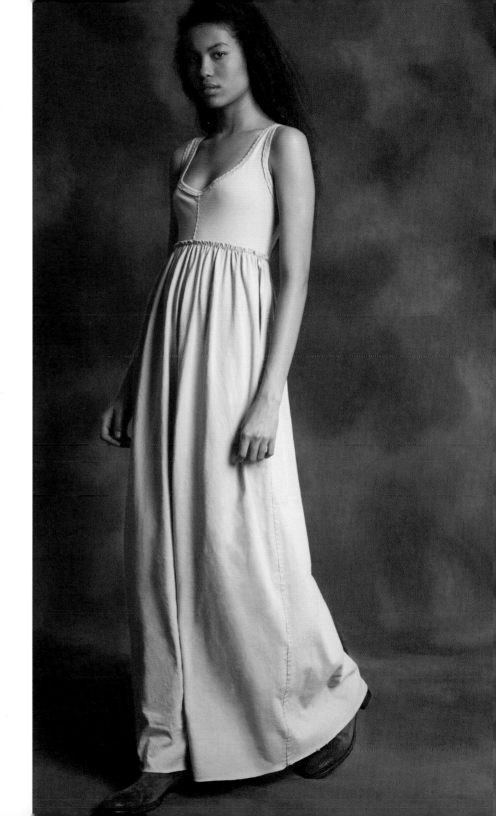

The single-layer Baby Doll Dresses shown here are made by following the instructions for the Baby Doll Top on page 57 and by appliquéing a 29"-wide ruffle to the hem of the constructed top for the short dress and a 45"-wide ruffle for the long dress. Shown at far right is our four-ruffle Peasant Dress.

Peasant Dress

The single-layer Peasant Dress at right is made by following the instructions for the ruffle border on page 57 and appliquéing an additional four rows of 7"-wide borders to the first ruffle, using the same gathering technique as for the ruffle border. Note that, as for the ruffle border, you'll need to cut the rectangle for the first border three times the length of the garment's hem edge and then cut each successive border three times the length of the previous border. You may also choose to vary this design by changing the size or number of the appliquéd borders. When only one ruffle is added, we call it a Baby Doll Dress.

Multiple Ruffle Borders

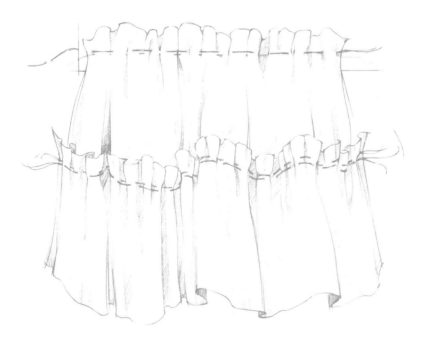

Cut first ruffle three times length of garment's hem edge; then cut each successive ruffle three times length of previous ruffle border.

Short/Mid-Length/Long Skirt

The directions below are for the Short Skirt but also apply to the Mid-Length Skirt and Long Skirt. Only the fabric yardage needed for each skirt is different (see pages 33 and 34 for the various yardages needed).

The instructions below can be used for either a single- or a double-layer garment.

Supplies

Short Skirt pattern (see pattern sheet at back of book)

1 yard of 60"-wide cotton-jersey fabric in one color, for top layer

1 yard of 60"-wide cotton-jersey fabric in second color, for backing layer (optional)

1 yard of 1½"-wide fold-over elastic

Paper scissors

Garment scissors

Tailor's chalk or disappearing-ink fabric pen

Hand-sewing needle

Button craft thread

All-purpose sewing thread

Pins

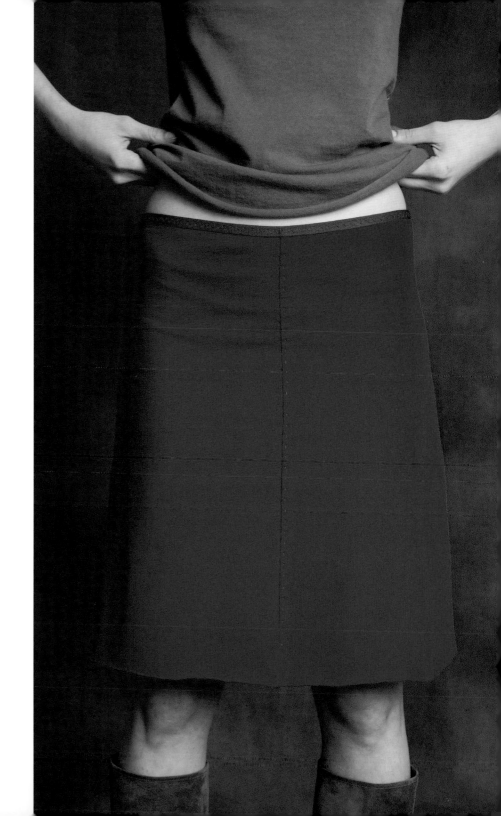

1. Prepare and Cut Pattern

Photocopy the Short Skirt pattern, and cut the photocopied pattern to your desired size, cutting as close as possible to the black cutting line. The Short Skirt pattern has two pattern pieces—a front panel and a back panel—with a ¼" seam allowance built into all the pattern edges.

2. Cut Top-Layer Pattern Pieces

Lay the top-layer fabric flat and fold to create two layers and position the front pattern on top of that yardage, making sure the pattern and fabric grain lines run in the same direction. With tailor's chalk, trace around the pattern's edges, remove the pattern, and cut out the traced pattern, cutting just inside the chalked line to remove it entirely. Repeat this step on the remaining top-layer fabric to cut four top layers for your skirt.

3. Cut Backing-Layer Pattern Pieces (for double-layer garment)

Lay out the backing layer fabric flat and fold to create two layers, and repeat the process in Step 2 to cut four backing-layer pieces.

4. Baste Waistline

To ensure that the waistline on your cut fabric pieces doesn't stretch while you're constructing your skirt, use a single strand of all-purpose thread to baste the waistline edges of each cut piece, as noted on the pattern piece.

5. Add Stenciling (optional)

Repeat Step 5 of the instructions for making the Bolero on page 50.

6. Pin Pattern's Top and Backing Layers (for double-layer garment)

Repeat Step 6 of the Bolero instructions.

7. Add Embellishment (optional)

Repeat Step 7 of the Bolero instructions.

8. Prepare for Construction

Once you've completed your embellishment, begin constructing your skirt by pinning each pair of adjacent panels with right sides together.

Shown at left is our single-layer Short Skirt. At right is our single-layer Mid-Length Skirt.

9. Construct Skirt

Thread your needle, love your thread, and knot off (see pages 21–22). Using a straight stitch, begin stitching the pinned pieces together, starting at the top edge of the skirt's waistline and stitching ¼" from the fabric's cut edges down to the bottom edge. Be sure to begin and end the seam by wrap-stitching its edges. Fell your seams by folding over each seam's allowances to one side and topstitching the seam allowances ⅛" from the cut edges, down the center of the seam allowances, using a straight stitch and wrap-stitching the seam. Repeat this process for the skirt's side seams.

10. Bind Waistline

Using 1½"-wide fold-over elastic (see page 8) and starting at the skirt's center-back waistline, encase the waistline's raw edge with the folded elastic, pinning or basting it in place as you work. Overlap the elastic's raw edges at the center back by about ½", and trim off any excess elastic. Using the stretch stitch of your choice, sew through all the layers down the middle of the elastic.

Shown at left is our single-layer Long Skirt.

Beaded Borders

It's easy to dress up a skirt, dress, or top with a beaded border. Shown here is our Bolero with long fluted sleeves with a 1" line of armor beading (see page 77) worked along the sleeve hems. The border accents the hem, adds a lovely shine, and gives some weight to it so that it drapes beautifully. When choosing stitches and beads for a border, I like to test the color combinations on an old T-shirt or other cotton-jersey garments to see if the combinations are appropriate. In one step, I have worked out my design choices and also embellished another piece in my wardrobe.

Basic Poncho

This useful accessory is one of my favorite pieces and a best seller for Alabama Chanin. Inspired by a poncho in the book *Weekend Knitting* (by my editor, Melanie Falick), it's quick and easy to make for oneself and for gifts. I keep one in my bag year-round to throw on as needed for warmth; all the women in my family do the same.

While beautiful in its simplicity as a single-layer, unembellished piece, this poncho is also stunning when embellished. Any of the embellishment techniques shown in this book can be used for this poncho, but those yielding the lightest-weight fabrics work best. For example, see the Negative Reverse Anna's Garden poncho on page 99. Techniques involving three layers of fabric with appliqués and other embellishments produce a more rigid fabric that doesn't drape as nicely as it should for a poncho.

This project requires 2 yards of fabric since the pattern's long side is cut with the fabric's grain. We've found that ponchos cut across the grain tend to stretch out of shape at the neckline over time (since fabric cut on the cross-grain stretches more and has less "memory" than fabric cut with the grain). In our studio, we always cut two ponchos at once by folding the fabric in half along the grain line to create a double layer and then cutting through both layers. If you want to follow our lead, a lucky friend or family member will thank you heartily.

Supplies

One 22" x 54" piece of pattern paper

2 yards of 60"-wide cotton-jersey fabric

Paper scissors

Garment scissors

Tailor's chalk or disappearing-ink fabric pen

Hand-sewing needle

Button craft thread

Pins

1. Prepare and Cut Pattern

On your 22" x 54" piece of pattern paper, draw a straight line parallel to the paper's long side, and label this line "Grain line." This is your Poncho pattern (see illustration on page 38).

2. Cut Out Poncho

Lay out your fabric either single-layer (if you want to cut one poncho) or double-layer (if you want to cut two). Place the Poncho pattern on top of the fabric, making sure that the pattern and fabric grain lines run in the same direction. With tailor's chalk, trace around the pattern's edges, remove the pattern, and cut out the traced pattern, cutting just inside the chalked line to remove it entirely.

3. Add Stenciling and Embellishment (optional)

If you want to embellish your poncho, cut any additional fabric layers called for by your chosen embellishment, stencil the top fabric layer, and let the stenciling dry completely. Then align and pin the top and any backing layers, and complete all embellishments that you want to add, stitching through all fabric layers you're working with. If you're adding beading to your project, avoid beading in the ¼" seam allowance.

4. Pin and Assemble Poncho

Lay out your cut single-layer or multi-layer embroidered poncho piece on a table, right side down. Bring the fabric's bottom right corner (A) up to meet the top left corner (C), and pin the rectangle's A-B edge to the C-D edge, starting at C and following the illustration on page 35. Thread your needle, love your thread, and knot off (see pages 21–22). Using a straight stitch, sew the pinned pieces together, starting at point A/C, stitching ¼" from the fabric's cut edges, and ending at point B/D. Be sure to begin and end the seam by wrap-stitching its edges to secure them. Leave the seam allowances floating.

5. Wash and Wear

We suggest washing the poncho before wearing it so that its raw edges curl.

Shown at left and right is our single-layer Basic Poncho. There is no front or back to this garment. Wear as desired, with the straight edge or V-shaped edge in front or back, or somewhere in between.

Tied Wrap

This wrap style has appeared and reappeared in fashion many times over the centuries—it's no wonder, given how simple and versatile it is! To wear the wrap, throw it over your shoulders and tie the two ends at the small of your back. I keep one of these wraps at the foot of my bed so that it's within reach if I feel chilly during the night. I also use the same version as a wrap for my daughter and have even tied one to a chair-back as a decoration.

At Alabama Chanin, we've embellished this wrap with a wide array of our favorite treatments, from couching (seen on page 110) to Alabama fur (see page 83).

Supplies

One 21" x 48" piece of pattern paper

1½ yards of 60"-wide cotton-jersey fabric

Tools for the stencil-transfer method of your choice (see page 17)

Paper scissors

Garment scissors

Tailor's chalk or disappearing-ink fabric pen

Hand-sewing needle

Button craft thread

Pins

1. Create Pattern

On the 21" x 48" piece of pattern paper, draw a straight line parallel to one of the paper's short sides, and label this line "Grain line." This is your Tied Wrap pattern (see page 38).

2. Cut Out Tied Wrap

Lay out your fabric double-layer and folded across the grain line, and place the Tied Wrap pattern on top of the fabric, making sure that the pattern and fabric grain lines run in the same direction. Using tailor's chalk, trace around the pattern's edges, remove the pattern, and cut out the traced pattern, cutting just inside the chalked line to remove it entirely. You'll now have a total of two pieces for your Tied Wrap.

If you're making this wrap as a basic garment without embellishment, cut a third layer to add more structure to the garment. If you plan to embellish your wrap, for example, like the Tied Wrap in Natalie's Dream on page 162 or the Alabama Fur Wrap on page 158, two layers of fabric is sufficient.

3. Add Stenciling (optional)

Although we didn't add stenciling to this basic version, you might want to add it to your Tied Wrap. If so, stencil the top layer of fabric now, and let the fabric dry thoroughly.

4. Align and Pin Fabric Layers

Align the cut backing and inner fabric layers, right side up, behind the top-layer fabric, also positioned right side up, making sure that all the fabrics' grain lines run in the same directions. Then pin the fabric layers together securely.

5. Add Embellishment (optional)

If you want to embellish your wrap, complete all the embellishments now, stitching through all fabric layers and working right up to all four edges of your Tied Wrap since the edges will be left raw and unseamed.

6. Add Ties and Wear

Cut two cotton-jersey ropes 18" long (see page 8), and attach each rope at the midpoint of each 21"-long edge with a couching wrap stitch (see page 111) ⅛" from the raw edge, as shown in illustration on page 35.

Shown at left and right is our single-layer Tied Wrap.

Bucket Hat

Many years ago a client bought several of these hats at one of our trunk shows in San Francisco and gave them to the famous guitarist Joe Satriani, who proceeded to wear them at performances around the globe. After seeing Joe perform in his multitude of hats, his fans began to email us asking for what they called the "Joe Satriani Hat." However, around here, we call it the Bucket Hat because one autumn we used one of these hats as a bucket to collect pecans. My daughter wears these hats, my son wears them, I wear them, in fact, we have so many different versions of them that it's difficult to tell how many we really own. However you decide to make your hat, make lots of them! Create one with a stitched border, as shown here, or make the more time-consuming couched (page 110) or Alabama fur version (page 82).

Supplies

Bucket Hat pattern (see pattern on page 39)

Cotton-jersey scrap or ½ yard of 60"-wide cotton-jersey fabric in one color

Cotton-jersey scrap or ½ yard of 60"-wide cotton-jersey fabric in second color

Paper scissors

Garment scissors

Rotary cutter and cutting mat

18" transparent plastic ruler

Tailor's chalk or disappearing-ink fabric-marking pen

Hand-sewing needle

Button craft thread

Pins

1. Create Pattern

Enlarge the Bucket Hat pattern on page 39 to your desired size on a photocopier, and use your paper scissors to cut out the photocopied pattern, cutting as close as possible to the black line. The Bucket Hat pattern has a total of two pattern pieces—a side and a top—with a ¼" seam allowance built into all the pattern edges.

2. Cut Top-Layer Pattern Pieces

Lay out your cotton-jersey scrap or yardage flat and folded in half on the grain. Lay the hat pattern's brim piece on the fold. Repeat the process to cut a second brim. Using tailor's chalk or a disappearing-ink fabric pen, trace around the pattern's edges, remove the pattern, and cut out the traced pattern piece, cutting just inside the chalked line to remove it entirely. Next lay the crown piece on the remaining folded fabric, making sure that the fabric and pattern grain lines run in the same direction, and cut out the crown. You'll now have two brim and two crown pieces, or four pieces, for your hat.

Lay out the cotton-jersey in the second color, flat and single-layer, to cut out the hat's contrasting band. Use the rotary cutter, cutting mat, and large plastic ruler to cut a 1"-wide strip across the grain to use for the contrasting band. Now you'll have a total of five cut pieces for your hat.

3. Add Stenciling (optional)

Although we didn't add stenciling to this basic version, you might want to add it to your hat's top fabric layer. If so, stencil the top layer now, and let the fabric dry thoroughly.

4. Pin Pattern's Top and Backing Layers

Align each cut top-layer piece on each corresponding backing-layer piece, with both fabrics facing right side up, and pat the layers into place so that their edges match. Then securely pin the two layers together on each piece. (Note that you'll add the contrasting band after constructing the hat.)

5. Add Embellishment (optional)

If you want to embellish your hat, complete all embellishments now, stitching through both fabric layers. If you're adding beading to your project, avoid beading in the ¼" seam allowance.

6. Prepare to Construct Hat

Once you've completed any embellishment you want to add to your hat pieces, begin constructing the hat by aligning and pinning the hat brim's short, straight edges, with right sides together, which will produce an inside floating seam (see page 45).

7. Construct Hat

Thread your needle, love your thread, and knot off (see pages 21–22). Using a straight stitch, sew the hat's pinned brim edges, sewing ¼" from the fabric's cut edges, and beginning and ending the seam by wrap-stitching its edges to secure them.

Next align and pin the hat crown's two layers, with both layers right side up; and then align and pin the crown to the brim circle, with the fabrics' right sides together. Using a straight stitch, sew the crown to the brim circle, being careful to keep your stitching tension even and tying off the thread after joining the two parts.

8. Add Contrasting Band

Pin your contrasting band in place along the outside layer of the hat's raw edge (so that the band will frame the face), overlapping the band's short edges by ⅛" to ¼" and positioning this overlap over the brim's side seam. Appliqué (see page 101) the band into place, using a row of herringbone stitches (see page 25) ¼" from the band's pinned raw edge and repeating the row of herringbone stitches on the band's other raw edge to secure it in place.

Fingerless Gloves

We received a pair of gloves like these—made from a recycled cashmere sweaters—as a thank-you gift from a woman named Debbie, who had attended two of our weekend workshops. As a thank you to her, we made a pair in our organic cotton-jersey. That's how one of our favorite accessories was born. The instructions below are for double-layer gloves.

Supplies

Fingerless Gloves pattern (see pattern on page 40)

Cotton-jersey scrap or ⅜ yard of 60"-wide cotton-jersey fabric in one color, for top layer

Cotton-jersey scrap or ⅜ yard of 60"-wide cotton-jersey fabric in a matching or second color, for backing layer

Paper scissors

Garment scissors

Tailor's chalk or disappearing-ink fabric-marking pen

Hand-sewing needle

Button craft thread

Pins

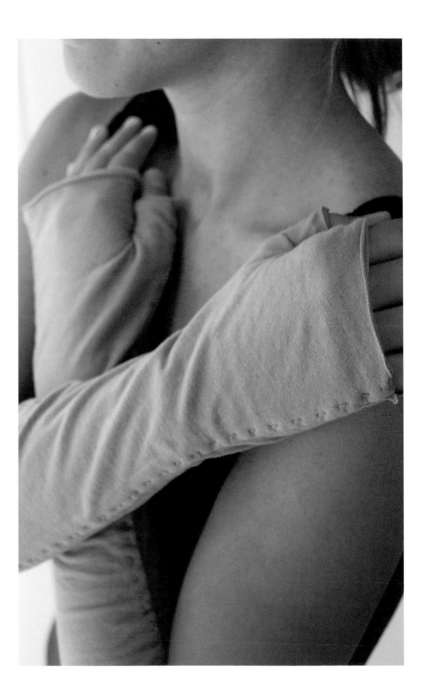

1. Create Pattern

Enlarge the Fingerless Gloves pattern on page 40 to your desired size on a photocopier, and use your paper scissors to cut out the photocopied pattern, cutting as close as possible to the black line. The Fingerless Gloves pattern has one pattern piece, and a ¼" seam allowance built into all the pattern edges.

2. Cut Top-Layer Pattern Pieces

Lay out your cotton-jersey scrap or yardage in the first color flat, and fold it to create a double layer. Lay the glove pattern on top of the fabric, making sure the pattern and fabric grain lines run in the same direction. Using tailor's chalk or a disappearing-ink fabric pen, trace around the pattern's edges, remove the pattern, and cut out the traced pattern pieces, cutting just inside the chalked line to remove it entirely. You'll now have two glove pieces. Repeat this process for a total of four glove pieces

Lay out the cotton-jersey scrap or yardage in the second color flat, and fold it to create a double layer. Position and cut the backing-layer pattern piece as you did for the top layer above. Repeat for the second glove for a total of eight cut pieces for your gloves.

3. Add Stenciling (optional)

Although we didn't add stenciling to this basic version, you might want to add it to your glove's top-fabric layer. If so, stencil the top layer now, and let the fabric dry thoroughly.

4. Pin Pattern's Top and Backing Layers

Align each cut top-layer piece on each corresponding backing-layer piece, with both fabrics facing right side up, and pat the layers into place so that their edges match. Then securely pin the layers together on each piece.

5. Add Embellishment (optional)

If you want to embellish your gloves, complete all embellishments now, stitching through both fabric layers. If you're adding beading to your project, avoid beading in the ¼" seam allowance.

6. Prepare to Construct Gloves

Once you've completed any embellishment you want to add to your glove pieces, begin constructing each glove by aligning and pinning the long, straight edges, with right sides together.

7. Construct Gloves

Thread your needle, love your thread, and knot off (see pages 21–22). Using a straight stitch, sew the glove's long, straight, pinned side edges, sewing ¼" from the fabric's cut edges and beginning and ending the seam by wrap-stitching its edges to secure them. Fell your seam by folding over the seam's allowance to one side and topstitching the seam allowance ¼" from the cut edges, down the center of the seam allowances, using a straight stitch and wrap-stitching the seam.

Next align and pin the long curved edge of the glove, with the fabric's right sides together and repeat the process above.

Repeat the process for the "V" that runs between the thumb and finger opening, and turn the glove inside out. Repeat the entire process for the second glove.

Chapter 5

Embellishing with Beads, Sequins + Embroidery

While our basic unadorned garments provide the structure for our couture collections, it's the embellishments that make them luxurious. Beads, sequins, and decorative stitches can transform a simple hand-sewn garment into an heirloom, even museum-quality, design.

Beading

Glass beads are available in a wide range of colors and finishes. We work with a selection of clear, opaque, and satin-finished beads. For seed, chop, and bugle beads, we choose the same button craft thread that we use for construction and embroidery, but use a single, rather than a double, thread since a double thread tends to tangle excessively when beading. When working with single thread, we secure it with at least two double knots (see page 21). We use either a #10 sharp or a #10 milliner's needle for beading because these needles accommodate our thicker thread while still passing through the beads' small holes. Always test a threaded needle with your beads to make sure they work well together before starting a project.

Beading Styles

We have developed a variety of beading styles with which to embellish garments. In each one, the beads are positioned on the fabric in a different way. When stitching beads to fabric, work the beads side by side, from one side of the shape you are covering to the other.

Full Beading describes an area completely or nearly completely covered.

Full Beading—Random Chop Beads

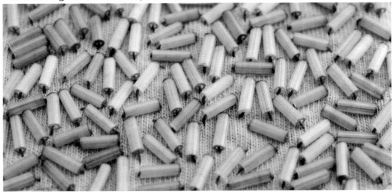
Full Beading—Random Bugle Beads

Full Beading—Random Mixed Beads

73

Half– (or Partial) Beading describes an area of fabric only partially filled with beads.

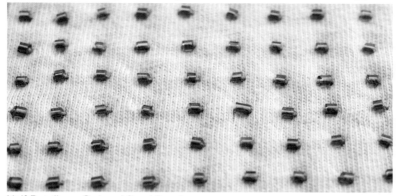

Half-Beading—Ordered Seed Beads

Half-Beading—Random Seed Beads

Half-Beading—Ordered Bugle Beads

Half-Beading—Random Bugle Beads

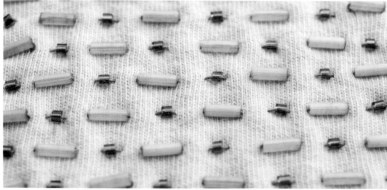

Half-Beading—Ordered Mixed Beads

Half-Beading—Random Mixed Beads

French Knot

The French knot is a beautiful way to add decoration to any fabric or project. While the French knot is made simply by knotting thread, it gives the appearance of beading, without the shine of the glass. You can also use French knots to secure sequins to a project, but do not use this knot to replace the double knot on page 21—French knots tend to come undone when not combined with a double knot.

French Knot

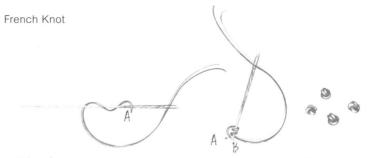

Bring thread up at A, and wrap twice around needle. Holding thread taut at base of needle, insert needle at B (as close as possible to A), and pull through to back of fabric to create knot on fabric's top surface.

French-Knot Patterning

An elaborate way to produce detailed patterning is to work groups of French knots within a stenciled pattern. The knots look like the random chop beads on page 73 but aren't shiny. We work French knots with a continuous thread, without double-knotting off each one individually, as long as the distance between the French knots does not exceed ¼".

The photo at left shows our Abbie's Flower stencil (see page 12) with French-knot patterning.

Sequins

Sequins, which are available in a wide range of shapes, sizes, and colors, are a beautiful way to highlight a stenciled pattern and add detail to a fabric. At Alabama Chanin, we usually use flat, round sequins in three different sizes: small (4mm), medium (5mm), and large (6mm). We sew sequins in place with a single strand of button craft thread, securing the thread with at least two double knots (see page 21). If we are placing sequins ¼" or closer together, we sew them with one continuous thread. If they are placed more than ¼" apart, we knot off the thread for each sequin separately.

French Knot Sequins

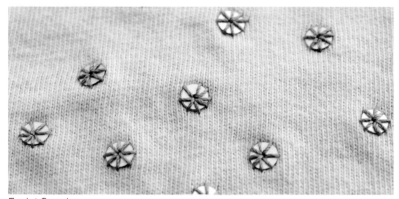

Eyelet Sequins

Seed Bead Sequins

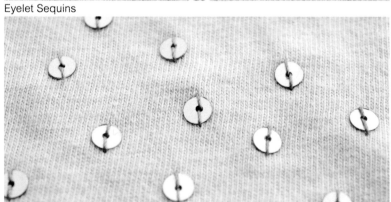

Straight Stitch Sequins

Chop Bead Sequins

Armor Beading: This technique combines chop bead sequins with random-mixed full beading (see page 73). Here the beading is gently fading from full beading to half-beading.

Decorative Stitches

The stitches featured in this chapter are deceptively simple yet produce visually complex results, and the possible variations are endless when worked in combination. Adding beads to a simple embroidery stitch creates an elaborate variation that can be used for necklines and borders, and for highlighting elements in a stenciled design. Decorative stitches can also be worked, with or without beads, in a circular fashion, as in eyelets and ermine on page 81 and Alabama fur on page 83. The techniques in this chapter require patience to master, but they are definitely worth the effort.

Beaded Stretch Stitches

While all of the stretch stitches from Chapter 3 have a specific technical purpose in construction, they can also be used as decorative stitches. Adding beads to each stitch dramatically changes its graphic quality, as you can see in the photographs at right. Instructions for making all of these stitches without beads are given starting on pages 25–27. Base the number of beads you pick up with each stitch on the desired effect.

Beaded Herringbone Stitch (see page 25)

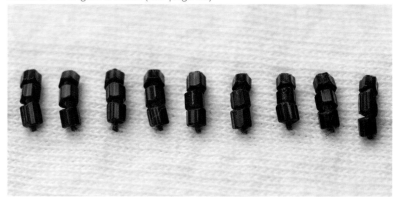

Beaded Parallel Whipstitch (see page 25)

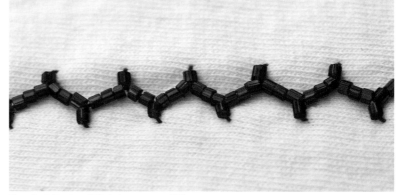

Beaded Cretan Stitch (see page 25)

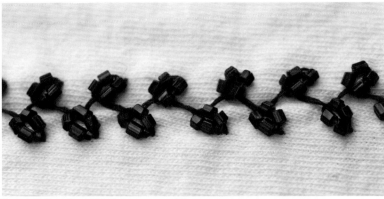

Beaded Rosebud Stitch (see page 25)

Beaded Zigzag Chain Stitch (see page 26)

Beaded Chevron Stitch (see page 26)

Beaded Featherstitch (see page 26)

Beaded Snail Trail Stitch (see page 26)

Beaded Cross-Stitch (see page 27)

Circular Stitches

There are a number of decorative stitches that are worked around a central point to create a circular pattern. These stitches can be combined in infinite ways to create beautiful fabrics like the Eyelets and Ermine fabric shown at left.

Ermine Stitch

Ermine stitch is a series of straight stitches worked around a central point to form a six-pointed (or more) star.

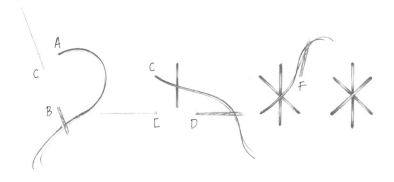

Bring needle up at A, go back down at B, and come up at C, making a long perpendicular stitch for ermine's center. Insert needle at D; come up at E, crossing over first perpendicular stitch; and go back down at F to complete cross.

Alabama Eyelets

At Alabama Chanin, we use the term *eyelet* for stitches worked in a circle around a point, although the term more typically denotes a small decorative or utilitarian hole. To create our eyelets, first, use a disappearing-ink fabric pen to draw a circle the size you want on your project's top layer, or use existing shapes in a stenciled design. Pin the top- and backing-layer fabrics together, making sure that the fabrics' grain lines run in the same direction. With a double strand of thread or four strands of embroidery floss, love your thread, knot off, and stitch around the shape, using a parallel whipstitch (see page 25) to make a whipped eyelet, a blanket stitch (see page 23) to make a blanket-stitch eyelet, or adding one or more beads to a straight stitch to create a beaded eyelet. Continue stitching around the circle until you arrive back at your starting point, and knot off the thread with a double knot on the same side of the fabric on which you originally inserted your needle.

Working Alabama Eyelets

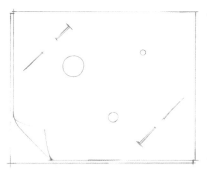

Using a disappearing-ink fabric pen, mark circles on top layer of fabric and pin to bottom layer of fabric.

Stitch around circles using stitch of choice.

Spirals, or Alabama Fur

One way to create a lushly textured fabric is to sew closely spaced, tiny spirals across the expanse of the fabric. The effect, which we call Alabama fur, is a combination of a backstitch (see page 23) worked in embroidery floss with our Spiral stencil (see page 86) and knots exposed on the right side (outside) of the fabric at the beginning and end of each spiral.

Like our couching treatment (see page 110), Alabama fur has been in continuous demand since we developed it a decade ago. See the beautiful Tied Wrap in Alabama fur on page 158, and the handsome Vitae Patterned Spiral Dress on page 20, which also includes detailed instructions for what we call the "double-stencil effect."

Supplies

Spiral stencil (see page 86)

Cotton-jersey fabric for top layer

Cotton-jersey fabric for backing layer

Textile paint

Spray bottle or airbrush gun, depending on stencil-transfer method (see page 17)

Embroidery scissors

Hand-sewing needle

Embroidery floss

Pins

1. Stencil Fabric
Stencil (see page 17) the fabric for the top layer, using our Spiral stencil; and set the fabric aside to dry thoroughly.

2. Pin Fabric Layers Together
Align and pin together the dried top layer and the bottom layer of fabric, with the fabrics' right side facing up and their grain lines running in the same direction.

3. Stitch the Spirals
Thread your needle with four strands of embroidery floss (or use two strands doubled), love your thread, and knot off with a double knot, leaving a 1" tail of floss. Note that this tail is slightly longer than we normally use when working with button craft thread.

Choose one of the spiral shapes in your stenciled design as a starting point; place your needle at the spiral's innermost point, directly on the stenciled line; insert the needle down through both layers of fabric; and bring it back up to the top layer ⅛" from your starting point, still on the stenciled-spiral line. Using a backstitch, sew along the stenciled spiral out to the end point, and tie off your embroidery floss on the top layer, using a double knot and again leaving a 1" tail of floss.

As you work, be sure to keep your stitches even in length and not too tight. Tight tension will cause your fabric to "shrink" substantially because Alabama fur has so many stitches per square inch.

Spirals with Inside Knots
Placing your knots on the inside (wrong side) of the fabric creates a beautifully textured fabric but eliminates the look of fur that outside knots produce.

Alabama Satin Stitch

In traditional embroidery, the individual stitches making up the satin stitch are worked as closely together as possible across the width of a shape to fill it. At Alabama Chanin, we work our satin stitch in a more spread-out fashion (leaving about $\frac{1}{16}$" between each stitch), following the outline of a stenciled shape but letting the textile paint or fabric background show through the threads. The beaded satin stitch is showcased beautifully in the Beaded Fern fabric on page 121.

Alabama Satin Stitch

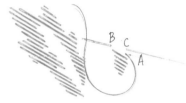

Work as for traditional satin stitch—bring needle up at A; go down at B, making straight stitch across shape; and come up at C—but spread out stitches about $\frac{1}{16}$" apart.

Stem Stitch

The stem stitch is one of our favorite stitches for defining small, thin lines, outlining shapes, and embroidering words, logos, or poetry onto our projects. The ropelike appearance of the stem stitch can be made thicker or thinner, depending on the angle at which you work your stitches. Use embroidery floss with the stem stitch to highlight delicate areas of your stenciled design.

Stem Stitch

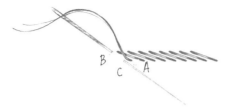

Bring needle up at A, slightly away from planned stitching line, and go back down at B, right on line, making a straight stitch that's angled slightly towards the line. Come up at C, at midpoint of previous stitch. Continue working stitches this way, following planned stitching line.

Spiral

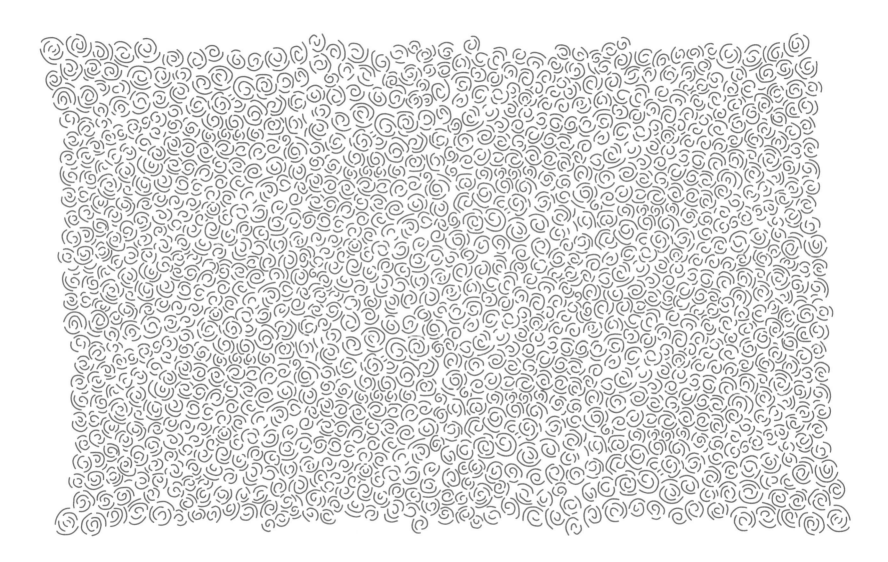

For the projects in this book, the Spiral stencil artwork was enlarged by 342 percent. This artwork can be photocopied and enlarged, or it can be downloaded full-size from www.alabamachanin.com.

Vitae

For the projects in this book, the Vitae stencil artwork was enlarged by 357 percent. This artwork can be photocopied and enlarged, or it can be downloaded full-size from www.alabamachanin.com.

Patterned Spirals, or Double-Stencil Effect

Recognizing that creating Alabama fur and spirals is time-consuming, we came up with patterned spirals, which requires working the spiral treatment only in selected areas rather than all over the fabric. This "selected-area" spiral treatment involves working with two stencils—the Spiral stencil and a second, different stencil—overlaying the pair so that the spiral pattern shows only through the cut-out areas of the second stencil. After securing the double stencils (we use paper clips to keep the two stencil layers in place) and positioning them on your fabric, apply textile paint through the stencils as usual to transfer the two composite designs.

When selecting stencils to combine for the double-stencil effect, it's important to choose two that contrast in scale—that is, pick one stencil with a small delicate design, like the Spiral stencil, and pair it with a second stencil with larger shapes, such as the Vitae stencil, to provide a framework for the smaller-scale design. This contrast in scale is important for patterned spirals since using two stencils of the same scale creates a totally different look.

Supplies

Spiral stencil (see page 86)

Vitae stencil (see page 87)

Cotton-jersey fabric for top layer

Cotton-jersey fabric for backing

Textile paint in one color

Textile paint in second color

Spray bottle or airbrush gun, depending on stencil-transfer method (see page 17)

Embroidery scissors

Hand-sewing needle

Embroidery floss

Pins

1. Prepare and Cut Out Cotton-Jersey Fabric
Cut out your cotton-jersey fabric in the color of your choice. We chose the same color for both the top and backing layers in the sample at left.

2. Stencil Design on Pattern Pieces
Lay out your top layer of fabric, right side up, and place the Vitae stencil on the fabric to begin stenciling. Using one color of textile paint and the stencil-transfer method of your choice, carefully transfer the Vitae stencil onto the fabric. Do not move the stencil, and let the paint dry to the touch. Then, with the dry Vitae stencil still in place, position the Spiral stencil on top of it (you can gently paper-clip or tape the stencils together, but it's not necessary if you work carefully), and transfer the spirals using a second color of paint, again letting the paint dry to the touch.

Reposition the Vitae stencil to the next area to be stenciled; transfer the stencil design, letting the paint dry to the touch, and repeat the above process with the Spiral stencil. Continue alternately transferring the Vitae and Spiral stencils until you've completely covered your project.

3. Stitch Your Patterned Spirals
Following the instructions for spirals on page 83, stitch your stenciled spirals inside each of the Vitae stencil shapes to create the patterned Alabama fur. Continue stitching until you've covered the entire project piece or fabric.

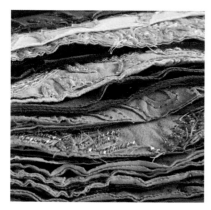

Sample Blocks & Library

When my first company, Project Alabama, closed in 2006 and my involvement with the label ceased, I took some time to think about what makes a good company better. I considered the mistakes we had made, what we had done well, who our customers were, and what they wanted. My goal was to incorporate these lessons into the new company—Alabama Chanin—we were building. It struck me that I had spent years developing incredibly detailed and elaborate fabric treatments for seasons that passed too rapidly. For all of my talk of recycling, sustainability, and placing value on the products we were producing, I had been devaluing our textile developments by discarding them every six months. I thought of all the fabrics, stencils, handwork, colors, testing, sewing, and re-sewing that had gone into developing these textile designs for which there was no documentation or swatches. When starting Alabama Chanin, I resolved to preserve the best of our developments in a library that we could return to over and over again. Without realizing it, I was also enriching our business by setting up a system for custom orders in which a client could become an active participant in the design process.

My simple decision to set up a standard way of recording the development of every design and fabric has led to an elaborate library consisting of more than five hundred swatches. To date we have twenty-seven books of fabric swatches that we use for everything from custom orders for clothing and interiors to inspiration for participants in our weekend workshops. When I spread these books across one of our 14'-long work tables, I am amazed to see the beauty we've created. In the past, we often tested colorways in garment form. Today, before any garment is cut, a tested and completed fabric sample swatch is approved. This small step has saved the company untold money, time, and waste.

I recommend that you start a library to document your own work. To do this, always create your samples at the same size so that your (master) pieces can be easily stored. And even if you don't want to keep the samples for posterity, you can use them to make pillows or patchwork quilts, or even frame a few to hang on the wall.

Chapter 6
Quilting + Reverse Appliqué

Featured in this chapter are two key techniques, quilting and reverse appliqué, both of which use the straight and backstitch to highlight stenciled shapes. Add to these techniques the beading and decorative stitches in Chapter 5 and the appliqué techniques in Chapter 7, and the possibilities are infinite.

Quilting

Traditional quilting uses small—even tiny—stitches to join multiple layers of fabric and/or batting. At Alabama Chanin, we stencil a pattern on a top layer of fabric, add a backing layer, and then straight stitch the top and backing layers together around the edge of each stenciled shape, using a longer stitch than the traditional quilting stitch.

Create the dress at left by following the instructions for our Fitted Dress on page 52, applying the Anna's Garden stencil, and reverse-appliquéing the stenciled design. For more information about this project, see our Index of Design Choices on page 164.

Reverse Appliqué

Reverse appliqué is worked on two layers of fabric: The top layer is stenciled and then stitched to the backing layer; next, part of the top layer is cut away to reveal the backing fabric underneath (see the photo at left). Below are instructions for the basic technique, followed by a dozen variations.

Reverse appliqué has defined my work for over a decade. Included in both *Alabama Stitch Book* and *Alabama Studio Style*, it is important to include here as well, as it is the basis for many of our fabrics.

Supplies

Stencil

Cotton-jersey fabric for top layer

Cotton-jersey fabric for backing layer

Textile paint

Spray bottle or airbrush gun, depending on stencil-transfer method

Embroidery scissors

Button craft thread

Pins

Needle

1. Transfer Stencil Design

Transfer the stencil design (see page 17) onto the right side of your top-layer fabric, and let the fabric dry thoroughly.

2. Pin Together Top-Layer and Backing Fabrics

Place the cut backing fabric, right side up, behind the area of the top-layer fabric to be appliquéd, making sure that the grain lines (see page 43) of both fabrics run in the same direction. Pin the two fabrics together securely.

3. Stitch Outline of Stenciled Shapes

Stitch the two layers of fabric together along the edge of each stenciled shape, using a straight stitch (see page 23). Be sure to love your thread (see page 22) and use a double knot (see page 21) at the beginning and end of each stenciled shape. Continue to move from one shape to the next, stitching around each one and always tying off with a double knot after completing each shape.

4. Trim Top Layer of Fabric Inside Stitched Shapes

Insert the tip of your embroidery scissors into the center of one of the stitched shapes, being careful to puncture only the top layer of fabric. Carefully trim away the inside of the shape, cutting about ⅛" from your stitched outline. This remaining ⅛" is wide enough to prevent the fabric and stitching from raveling, and narrow enough to display the reverse appliqué pattern nicely (along with a sliver of the original stenciled design's paint color). Never cut closer than ⅛" to any stitched outline.

After trimming the top layer of fabric on every shape possible—very small elements in the stenciled design may be too small to trim, so leave them uncut—you have two options for finishing the backing fabric: Either leave the backing fabric as is; or turn the project wrong side out, and trim away the backing fabric around each stitched shape, leaving a ⅛" border outside the stitched outline. Use extreme caution when using the second method to avoid making holes in your fabric by cutting in the wrong place.

Reverse Appliqué with Bugle-Beaded Stitches

You can add a bugle bead to each of your straight stitches to enhance the reverse appliqué.

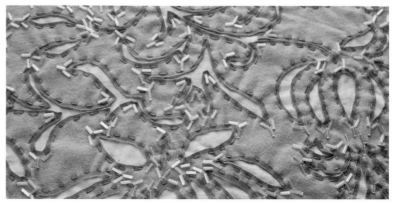

Accent-Beaded Reverse Appliqué

We often use accent-beaded reverse appliqué to visually join individual stenciled shapes. The beading is done freehand—that is, without a stenciled shape to follow—making this a favorite among those who love "accidental" design. In the fabric shown here, we used accent beading to create stems and leaves and unify the floral motif.

Reverse Appliqué with Chop-Beaded Stitches

Alternatively, you might want to add three chop beads to each straight stitch to create a more heavily beaded effect.

Inside-Beaded Reverse Appliqué

Although it is time-consuming, inside-beaded reverse appliqué produces a stunning effect. The beads add substantial weight and structure to the cotton-jersey fabric, creating garments that drape beautifully. Follow the instructions for reverse appliqué on page 95, and add beading style of your choice inside the area you trim away to reveal the backing fabric, leaving the ⅛" sliver of the stenciled top layer visible.

Reverse Appliqué with Inside and Accent Beading

The combination of inside and accent beading adds to the depth and richness of any pattern.

Backstitched Reverse Appliqué

This technique uses the backstitch rather than the traditional straight stitch. You'll find that fabric worked with this technique is equally intricate and lovely on both the right and wrong sides. Embroidery floss is not a good choice for basic reverse appliqué since it's not strong enough, but it works well for backstitched reverse appliqué since the extra stitches making up the backstitch provide the additional strength needed.

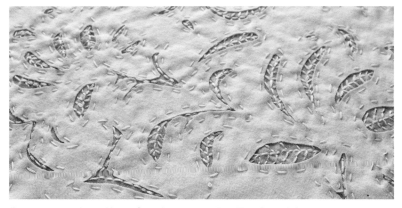

Reverse Appliqué with Inside Embroidery

This technique is similar to reverse appliqué with inside beading at left, in that it adds a decorative element inside a cut reverse-appliqué shape. My favorite way to work this technique is with the stem stitch (see page 85) since you can actually build the visual structure, or "bones," of a stenciled shape with your thread. I like to use embroidery floss for inside embroidery because of its light sheen and the wide variety of colors it comes in.

Bugle-Beaded, Backstitched Reverse Appliqué

This technique is worked the same as backstitched reverse appliqué, with a bugle bead or a number of seed beads added to each backstitch. For this technique, I recommend using our standard double strand of button craft thread rather than embroidery floss, which will not be strong enough to support the beads' extra weight.

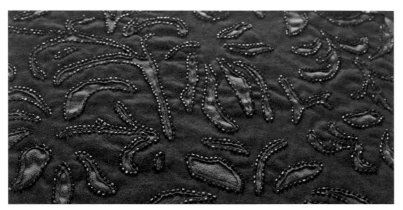

Seed-Beaded, Backstitched Reverse Appliqué

For this effect, add three seed beads to each of your backstitches.

Inked-&-Beaded, Backstitched Reverse Appliqué with Inside Beading

After using an ultra-fine-point permanent marker to trace the outside edge of your stenciled design, stitch a beaded-backstitch line ⅛" inside your stenciled shape; then cut away the top layer of fabric ⅛" inside your beaded, backstitched line. Finally, fill the exposed backing fabric with inside beading.

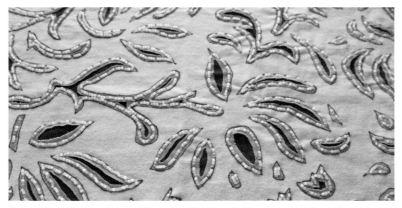

Inked-&-Beaded, Backstitched Reverse Appliqué

This technique adds a graphic quality and a lot of depth to the surface of a fabric. After stenciling your fabric, use an ultra-fine-point permanent marker to trace the outside edge of each stenciled shape. Then sew a beaded backstitch ⅛" inside the edge of the stenciled shape; and finally cut away the top layer of fabric ⅛" inside the edge of the beaded-backstitched line, revealing the backing fabric underneath.

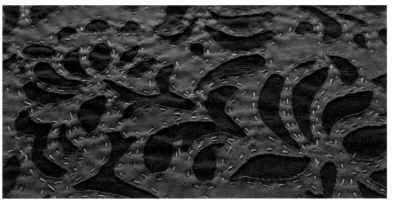

Outside Reverse Appliqué

To create outside reverse appliqué, follow the same steps as for reverse appliqué on page 95, but place the stitches outlining the stenciled shape ⅛" outside of the stenciled line. Then cut the top layer of fabric on the stenciled line, removing all textile paint. I like to use this technique with top-layer and backing-fabric colors that are close in tone to create a delicate shaded effect.

Negative Reverse Appliqué

Negative reverse appliqué looks like appliqué but is worked as reverse appliqué: After stenciling your top-layer fabric and placing it on top of your backing fabric, straight-stitch ⅛" inside the edge of the stenciled shape; then cut the top layer of fabric ⅛" outside the edge of the stenciled shape, leaving a ¼" sliver of top-layer fabric beyond your stitching line.

Create the poncho above by following the instructions on page 64 and embellishing with negative reverse appliqué Anna's Garden. For more information about this project, see our Index of Design Choices on page 164.

Chapter 7
Appliqué

Appliqué is a way of "applying" one fabric on top of another fabric. At Alabama Chanin, we use appliqué to add color, texture, and dimension to our work. Appliqués can be stitched to the base fabric with a variety of stitches, both simple and decorative, producing stunning effects. We generally use a stencil to transfer a design to both the base fabric and appliqué fabric, but you can also draw or paint and cut freehand any shape you want to appliqué.

Supplies

Stencil

Cotton-jersey fabric for top layer

Cotton-jersey fabric for appliqué pieces

Permanent marker or textile spray paint

Spray bottle or airbrush gun, depending on stencil-transfer method (see page 17)

Embroidery scissors

Button craft thread

Needle

Pins

1. Stencil Pattern on Base Fabric

Stencil (see page 17) a pattern on the right side of your base fabric where you want to stitch the appliqué pieces, remove the stencil, and let the fabric and stencil dry thoroughly.

2. Cut Out Appliqué Pieces

To make your appliqué pieces, flip the dried stencil used in Step 1 to the wrong side, and transfer the stencil pattern to the wrong (back side) of the appliqué fabric. After letting the stenciled fabric dry, begin by cutting out one stenciled shape, ¹⁄₁₆" around the outside of the stenciled edge. Once you cut out the shape, flip it over, right side up, and pin it to the corresponding shape in the stenciled pattern on the base fabric. Repeat for your entire stenciled design by cutting one piece at a time and pinning it into place.

3. Stitch Appliqué Pieces to Project

Position each cut appliqué shape, right side up, on top of the corresponding shape in the stenciled design on the base fabric. It's important to match up each shape as you cut it—unless you're fond of jigsaw puzzles! Align the edges of the appliqué and stenciled shape, pin the appliqué securely in place, and attach the appliqué's raw cut edges using the parallel whipstich (see page 25). The straight stitch (see bottom right) is the easiest to use, while the parallel whipstitch (see top right), which secures the fabric extremely well, is the stitch we use most often at Alabama Chanin.

Appliqué with Beaded Parallel Whipstitch

Adding beads to each parallel whipstitch produces a beautiful hatching effect. For the fabric above, we used bugle-beaded parallel whipstitches for the larger stenciled shapes and one or two chop beads on the whipstitches for the smaller stencil shapes.

Appliqué with Straight Stitch

This technique adds a "rawer" look to the fabric than does appliqué attached with parallel whipstitch. Align the cut appliqué piece, right side up, on the stenciled shape on your top-layer fabric; and stitch ⅛" inside the edge of the appliqué. After you wash the fabric, the appliqué's edges will curl slightly, adding more dimension to the fabric.

Appliqué with Blanket Stitch

I love the unified appearance of this technique, which involves attaching appliqué to a base fabric with a blanket stitch. The resulting appliqué has a stronger visual presence and blends less with the base fabric than when applied with the more common parallel whipstitch.

Appliqué with Backstitch

When appliquéing fabric with a backstitch, we prefer to use embroidery floss because of the range of floss colors available. You might also decide to add additional rows of backstitches in a range of coordinating colors or sizes for a rich, varied design.

Appliqué with Beaded Straight Stitch

You can add a bead to each straight stitch while attaching your appliqué to create a more defined outline for the appliquéd shape. In the photo above, we have added three chop beads to every stitch.

Appliqué with Beaded Backstitch

This technique combines appliqué with continuous lines of beads and is perfect for eveningwear. As with the reverse appliqué with beaded backstitch, we recommend using our standard double strand of button craft thread rather than embroidery floss because the floss is not strong enough to support the beads' extra weight.

Relief Appliqué

The process for creating relief appliqué is the same as for creating our standard appliqué (see page 101), except that the appliqué pieces are cut about 15% larger than the stenciled design to which you'll appliqué them. Forcing the larger appliqué piece into a smaller area stenciled on the base fabric naturally wrinkles the appliqué, and these wrinkles remain intact after you iron the completed appliqué. This technique requires preparing two sets of stencils: one for the base fabric and a second one that's 15% larger for the appliqué. The appliqué pieces are attached to the base fabric with parallel whipstitch.

Beaded Relief Appliqué

Add bugle, seed, and/or chop beads to each of your parallel whipstitches to give additional dimension to your relief appliqué. In the fabric above, a bugle bead has been added to each of the parallel whipstitches in the larger parts of the stencil, and one to two chop beads have been added to the whipstitches in the smaller parts of the stencil.

Cretan Stripes

A basic stripe inspired several of our embellishment techniques, including the Cretan stripes shown here. Many variations can be achieved by varying the size of the stripe, the stripe color, and/or the stitch you use to secure the stripe to the fabric. A collection of our favorite variations is presented here.

Supplies

Cotton-jersey fabric for top layer

Cotton-jersey fabric for backing layer

Cotton-jersey fabric for appliqué

Garment scissors

Rotary cutter and cutting mat

18" transparent plastic ruler

Tailor's chalk or disappearing-ink fabric pen

Hand-sewing needle

Button craft thread

Pins

1. Draw Lines on Base Fabric

Using tailor's chalk or a disappearing-ink fabric pen, draw lines 2" apart on the right side of your base fabric to create a design, or sewing guide, for where you want to stitch your appliqué stripes.

2. Cut Out Appliqué Pieces

Cut ¾"-wide cotton-jersey strips (for stripes) across the fabric's grain (see page 43) and as long as you need to complete your panels. (If you find that you need longer appliqué pieces, simply overlap the ends of your appliqué stripes by ¼", and treat the two pieces as one.)

3. Stitch Appliqué Pieces to Project

Position each cut appliqué stripe, right side up, on the base fabric, using the drawn lines as a guide; and pin the stripes securely in place. Appliqué the stripes to the base fabric using a Cretan stitch (see page 25) down the middle of the stripe. The finished stripes will be 1¼" apart.

As a variation, you might also try changing the colors, widths, and position of the stripes to create a very textured landscape on the fabric's surface.

Stripe with Chain Stitch

Create additional texture by attaching a ½"-wide stripe of fabric using a straight-line chain stitch down the middle.

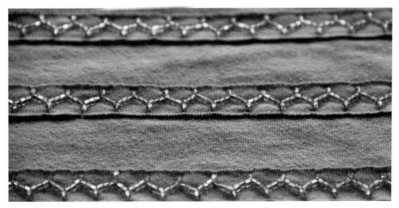

Beaded Appliqué Stripe

Add beads to your decorative stitch when applying stripes to fabric to create a richly textured surface for an eveningwear project.

Stripe with Beaded Chain Stitch

For an elegant variation, add beads to the straight-line chain stitches used to appliqué stripes. The Kristina's Rose fabric on page 126 shows this variation worked in a circular fashion.

Random Ruffle

We developed this technique in 2001 for our second collection of T-shirts. The ruffle can be used to embellish existing pieces of clothing and to create a wonderful tuxedo-like Fitted Top like that on page 159.

Supplies

Cotton-jersey fabric for project

Cotton-jersey fabric for appliqué pieces

Rotary cutter and cutting mat

Tailor's chalk or disappearing-ink fabric pen

Embroidery scissors

Button craft thread

Needle

Pins

Beaded Random Ruffle

For this variation, randomly add bugle and chop beads to individual stitches across the entire ruffle.

1. Draw Lines on Base Fabric

Using tailor's chalk or a disappearing-ink fabric pen, draw a line—or several lines—on the right side of your base fabric to create a design, or guide, for where you want to add random ruffle stripes.

2. Cut Out Appliqué Pieces

Cut ½"-wide cotton-jersey strips (for stripes), cutting across the fabric's grain (note that this is different from the instructions for Cretan stripe on page 104) as long as you need to complete your panels. (If you need longer appliqué pieces, simply overlap the ends of the appliqué stripe by ¼" and treat the two pieces as one.)

3. Stitch Appliqué Pieces to Project

Position an appliqué stripe, right side up, on the base fabric, using one of the drawn lines as a guide. and pin the beginning of the stripe securely in place. Insert a needle in the middle of the stripe and ⅛" from its beginning edge, and sew one stitch to anchor the stripe, followed by a couple more straight stitches. Then, using your fingers, "ruffle" your stripe by gathering 1" to 1½" of it onto your needle, and make one stitch into the base fabric. Sew several straight stitches in the stripe and base fabric, and repeat the ruffling process, varying the size and position of each gathered ruffle and straight stitches to produce a "random ruffle." Keep working this way until you get to the end of the drawn line, and tie off your thread. Repeat this process on each of your drawn lines.

Random Ruffle

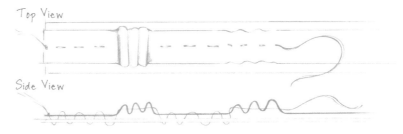

To make a random ruffle, work a few straight stitches into stripe and base fabric, and then work a few stitches only into stripe fabric, gathering loosely.

Folded-Stripe Appliqué

For this variation, hold two ½"-wide strips of fabric together, then randomly fold them back and forth along an imaginary line while sewing them into place with a straight stitch. Folding the stripes creates a relief effect, giving real dimension to the finished project. We like to use tonal combinations of colors to enhance the effect.

Ruffle Stripe

To make a ruffle stripe, cut a 1"-wide strip of cotton jersey, sew a basting stitch down the middle of the strip, and pull on the ends of the basting thread to ruffle, or gather, the strip. Attach the ruffled stripes to your fabric by first basting them down with an all-purpose thread and then securing them with a stretch stitch or another decorative stitch down the middle of the ruffle.

Folded-Stripe Appliqué with Beaded Straight Stitch

For more drama, add a bead to each of the straight stitches while attaching your folded appliqué to the base fabric.

Beaded Ruffle Stripe

Add beads to your ruffle stripe for extra dimension and detail.

Pleated Ruffle

A pleated ruffle is similar to a ruffle stripe (see opposite page) but looks more orderly and defined. To make a pleated ruffle, first cut 2"-wide strips of fabric across the grain. Pleat each fabric strip with 1"-wide folds—so ½" of fabric shows—basting each pleat in place down the center as you fold it. Using tailor's chalk or a disappearing-ink fabric pen, draw a line on the top layer of your fabric where you want to position the pleated ruffle, and baste the pleated ruffle in place. Cut a ½"-wide strip of cotton jersey in a contrasting color, and apply it to the middle of your pleated ruffle, using a decorative or stretch stitch.

Pleated Ruffle with Beaded Stitch
Add one or more seed or chop beads to each of your appliqué stitches to lend a more elegant appearance to this fabric.

Couching

Because of its sculptural quality, couching is one of our customers' favorite techniques. We use it on coats, dresses, skirts, tops, pillows, blankets, and an array of other pieces for the body and home.

Traditional couching is a very old embroidery technique, in which yarn (or another material) is laid across the surface of a ground fabric and sewn into place, often with a satin stitch. At Alabama Chanin, we use the term *couching* to describe a form of appliqué in which cotton-jersey ropes (see page 8) are appliquéd to a base fabric with a parallel whipstitch, following a design stenciled on the fabric.

While couching is a simple concept, it's one of our more advanced techniques. Because it's nearly impossible to pin the narrow rope to the fabric before sewing it in place, you must use your fingers to hold—or "sculpt"—it into position as you go.

Supplies

Stencil

Cotton-jersey fabric for top layer

Cotton-jersey fabric for backing layer

Cotton-jersey fabric for ropes

18" transparent plastic ruler

Rotary cutter and cutting mat

Textile paint

Spray bottle or airbrush gun

Embroidery scissors

Hand-sewing needle

Button craft thread

Pins

1. Stencil Fabric and Prepare Ropes

Stencil the right side of your top-layer fabric, and set it aside to dry thoroughly. Using the fabric for your ropes, cut ½"-wide strips, cutting them with the grain and making them as long as you want. Pull each strip from both ends at the same time to make ropes about ³⁄₁₆" in diameter.

2. Align Top and Bottom Fabric Layers

Align the top and bottom layers of fabric, both right side up and with the grain lines running in the same direction, then pin the two layers together.

3. Prepare for Couching

Thread your needle with a double length of thread, love your thread (see page 22), and knot off (see page 21) with a double knot. Choose one shape in your stenciled design as a starting point. Place one end of a cotton-jersey rope at the edge of that stenciled shape, leaving about ½" of rope free beyond that point; insert your needle from the back of the fabric up through the middle of the rope to secure it with a couching wrap stitch (see below) at the edge of the stenciled shape, bringing the needle back down through both pinned layers of fabric to prepare for the next step.

4. Couch First Stenciled Shape

Using your fingers, hold the secured rope along edge of stenciled shape, and work one couching stitch around rope to anchor it in place by bringing needle up at A and going back down at B through both layers of fabric. Realign rope with next part of stencil design's edge, sew next couching stitch about ⅛" to ¼" away, and continue this process around this stenciled shape to arrive back at your starting point.

Couching Stitch

5. Finish Couching First Stenciled Shape

Trim the cotton-jersey rope so it overlaps the beginning end by ⅛", and secure the overlapped ends with a couching wrap stitch, stabbing the needle through the ends and pulling the thread through to the back of the work. Knot off your thread using a double knot.

Beaded Couching

Incorporate extra detail into a couched stenciled pattern by adding one or more seed or chop beads to each stitch.

Cotton Yarn Couching

For a more detailed effect, work couching with cotton yarn.

Latticework Passementerie

Passementerie is the French term for (often handmade) ornamental edging or trim made from braided yarns, cords, and threads. When I lived in Vienna in the 1990s, there were wonderful old-fashioned shops filled with the most beautiful trims and braids I have ever seen in my life. Alabama Chanin passementerie is inspired by the days that I spent in those shops. Our version of passementerie is really a form of couching with our cotton-jersey ropes.

The passementerie in the photo at right was done with parallel lines drawn 3" apart across the fabric's grain and used as a stitching guide. This technique works equally well for borders or any other place you want to add a decorative detail.

Supplies

Cotton-jersey fabric for top layer

Cotton-jersey fabric for backing layer

Cotton-jersey fabric for ropes

18" transparent plastic ruler

Rotary cutter and cutting mat

Tailor's chalk or disappearing-ink fabric pen

Embroidery scissors

Hand-sewing needle

Button craft thread

Pins

1. Draw Lines on Fabric
Using tailor's chalk or a disappearing-ink fabric pen, draw two parallel lines where you want to place the passementerie.

2. Make Cotton-Jersey Ropes
Cut cotton-jersey strips ½"–1½" wide (depending on effect desired), and simultaneously pull both ends of each strip to create ropes.

3. Pin Top and Bottom Layers Together
Align and pin together the top and bottom layers of fabric, making sure the fabrics' grain lines run in the same direction.

4. Pin Ropes in Place
Pin the end of one cotton-jersey rope at the beginning of one drawn line, with the rope's additional length laid in the direction of the opposite line. Pin one end of a second cotton jersey rope to the beginning of the other line, with its additional length laid in the direction of the first line, creating an X between the two lines. At 1" intervals, pin each cotton-jersey rope to the opposite line, overlapping the two ropes as they cross in the middle.

5. Attach Ropes and Finish Passementerie
Thread two needles with button craft thread, love your thread, and knot off. Begin sewing with one needle by using the couching wrap stitch (see page 111) to anchor one end of the pinned rope to the edge of your top line. Then use a decorative stitch (see pages 25-27) to secure the top edge of the pinned rope, sewing across all the loops in the passementerie ⅛" below their top turn. When you finish attaching the passementerie's top edge, secure the cut end of the rope with a couching wrap-stitch. Then use the second threaded needle to repeat the stitching process to attach and finish the bottom edge of all the loops in the passementerie.

If you run out of rope as you work across your project, simply overlap the end of one rope and the beginning of the next rope by ⅛" and continue sewing.

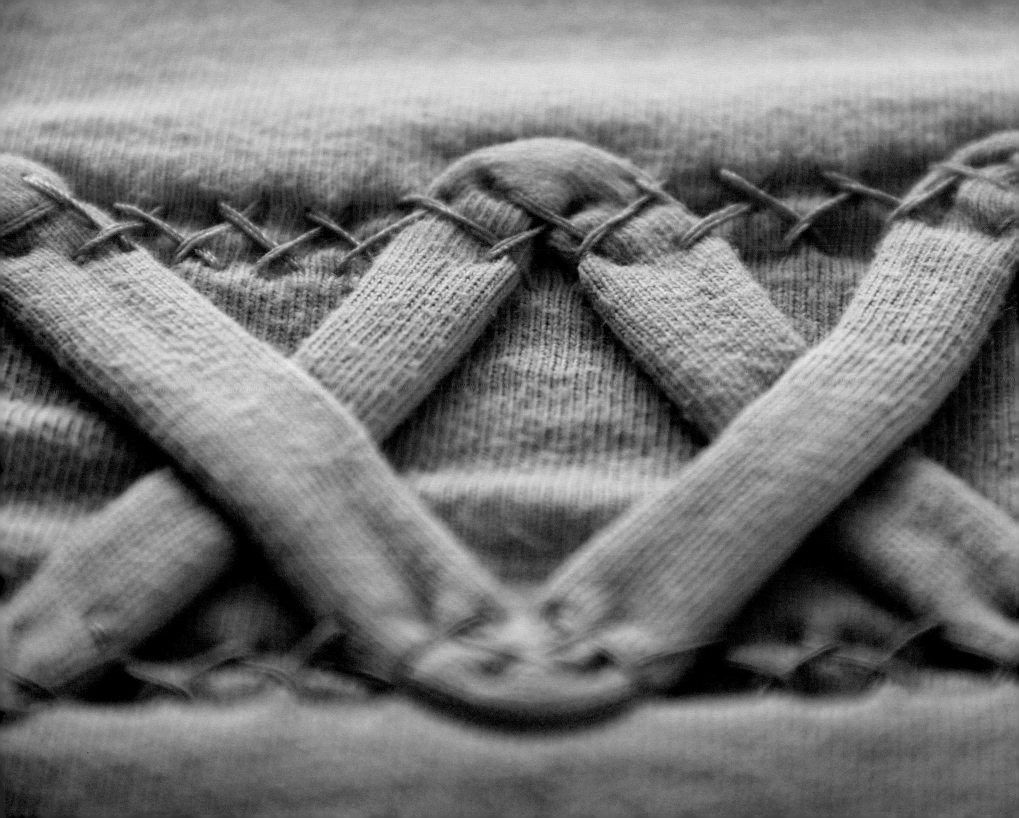

Ruffle Passementerie

Ruffle passementerie is similar to latticework passementerie (see page 112) in that it's constructed using looped ropes of cotton jersey; but, in this technique, the passementerie is worked along one line with ropes of varying sizes and loops of varying widths. The photo at left shows passementerie made in two steps: First, a cotton-jersey rope, whose original cut width was 1", was looped in a 2"-wide figure-8 pattern pattern across the length of the line and sewn down with a straight stitch. Second, another rope, whose original cut width was ½", was looped in a 1½"-wide snake pattern on top of the first row, with each loop secured with one small stitch, working from loop to loop in zigzag fashion.

Create the scarf at right by applying ruffle passementerie down the middle of a 7" x 36" rectangle of double-layer cotton jersey on both sides.

June's Spring

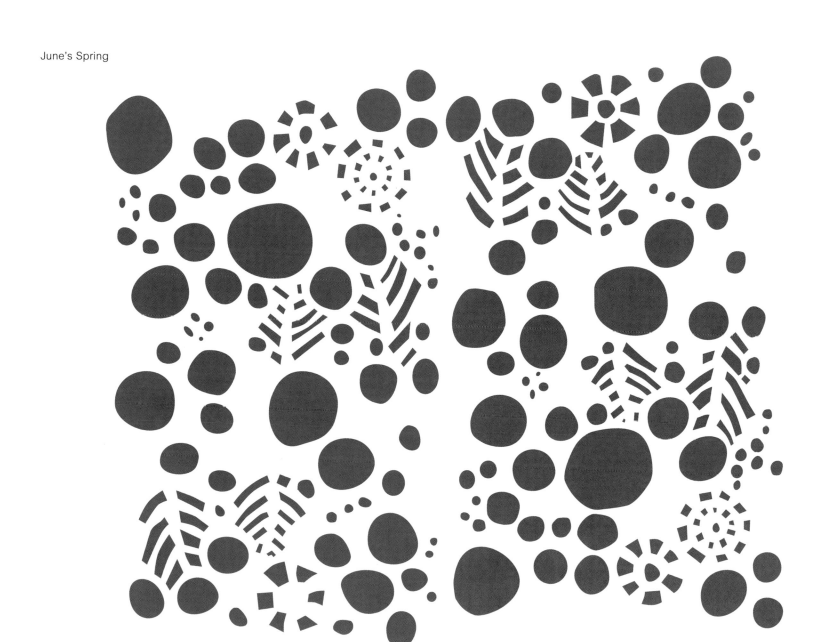

To create the pieces shown in this book, June's Spring was enlarged by 300 percent. This artwork can be photocopied and enlarged, or it can be downloaded full-size from www.alabamachanin.com.

Chapter 8
Fabrics + Fabric Maps

In this chapter are close-up photos of seven beautiful fabrics along with the stencils on which they are based. Also included are detailed "maps" showing exactly how to create the fabrics, including the placement of every stitch, applique, and bead.

We start with June's Spring, a fabric inspired by the Arts & Crafts fabrics of William Morris and named for one of our colleagues. Beaded Fern on pages 120 and 121 has a particularly organic but elegant feel and is among our clients' favorites. It is followed by Paisley Delight on page 122. The paisley—a tear-, pear-, or kidney-shaped curved figure—is a common motif in almost all cultures across the globe. Ours includes leaf and flower motifs as well as lots of beading. Kristina's Rose fabric on page 126 uses stripe appliqué techniques from Chapter 7 in combination with a beaded stitch from Chapter 5, all worked in loose, undulating circles.

Next come Satin Stars on pages 128 and 129 and Climbing Daisy on page 130. Satin Stars originated as a way to add a "punk rock" detailing to T-shirts but quickly migrated to dresses, skirts, and coats. Climbing Daisy is our homage to ribbon embroidery, which uses ribbon rather than thread or embroidery floss to stitch a design. You can actually use ribbon embroidery for any of the stitches in this book.

The final fabric in this chapter is Natalie's Dream, which feels like a culmination of my dreams in textile design. It is one of my all-time favorite fabrics. You can incorporate these fabrics into patchworks, full garments, or simple swatches intended for framing. However you choose to use them, I hope that they will become jumping-off points for your own unique dreams.

June's Spring Fabric

To create June's Spring: **1.** Stencil fabric using June's Spring stencil from page 116. **2.** Backstitch circles by stitching directly on the stenciled edge. **3.** Reverse-appliqué some of the stitched circles. **4.** Cut circles in two different colors for appliqué. **5.** Appliqué circles to fabric base with a whipstitch. **6.** Backstitch one of the tree shapes, and add

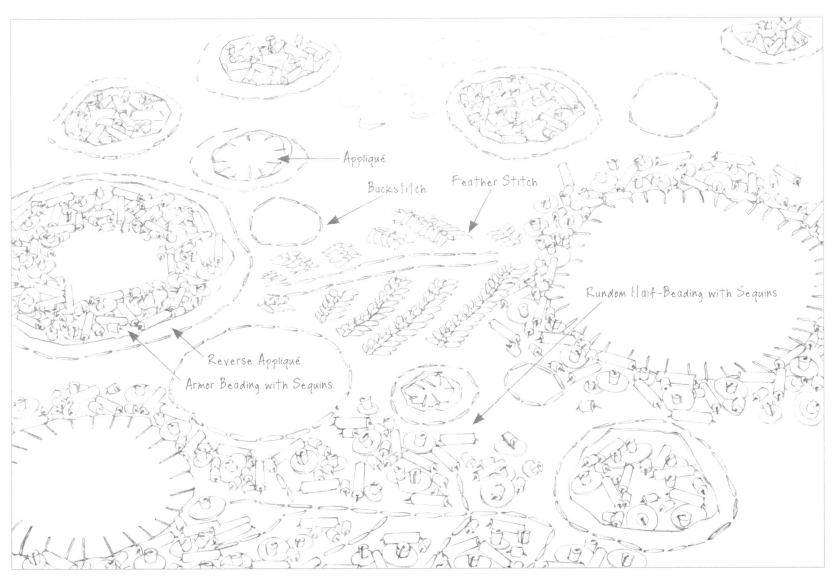

Appliqué

Backstitch

Feather Stitch

Random Half-Beading with Sequins

Reverse Appliqué
Armor Beading with Sequins

random half-beading (see page 73) with sequins (see page 76). **7.** Choose another tree shape, backstitch up the middle, and add Feather stitch (see page 26) to each branch with embroidery floss. **8.** Add random half and full-beading (see page 74) to the areas indicated. **9.** Continue to work fabric until all of the stenciled shapes are completed.

Fern

To create the projects in this book, Fern was enlarged by 285 percent. This artwork can be photocopied and enlarged, or it can be downloaded full-size from www.alabamachanin.com.

Beaded Fern Fabric

Chop or Seed Bead

Satin Stitch

To create Beaded Fern: 1. Transfer the Fern stencil at left to your fabric or project. **2.** Using a double length of button craft thread, sew a satin stitch through all fabric layers, across the leaf's stenciled shape, placing your stitches about 1/16" apart. On every other stitch, pick up one chop or seed bead.

3. Continue working the beaded fern stitch along the leaf shape, alternating between beaded and unbeaded stitches until you reach the end of the stenciled shape. **4.** Tie off the thread on the fabric's top layer, leaving long thread tails. **5.** Repeat this process until you have stitched each of your stenciled shapes.

Paisley Delight Fabric

To create Paisley Delight: 1. Stencil fabric using the Paisley stencil from page 124. **2.** Backstitch-reverse-appliqué all of the stenciled shapes by stitching directly on the stenciled lines and cutting away the inside of the stitched shapes. **3.** Use armor beading technique (see page 77) to fill in each of the cut-away stenciled shapes

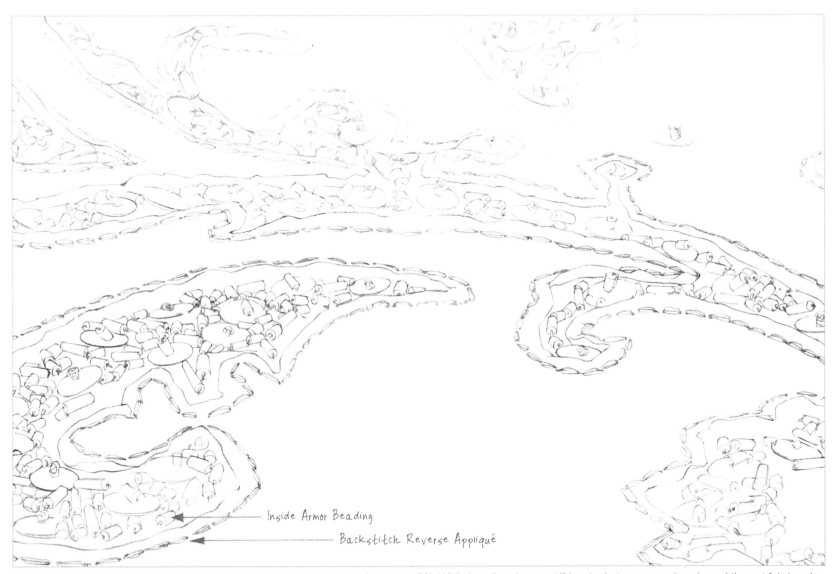

Inside Armor Beading

Backstitch Reverse Appliqué

with bugle beads, chop beads, seed beads, and seed-beaded sequins (see page 76). While beading, leave a ⅛" border between your beads and the cut fabric edge to create a stripe of color around your beadwork. **4.** Continue with this method until you have filled in all your backstitched-reverse-appliquéd stenciled shapes.

To create the projects in this book, Paisley was enlarged by 370 percent. This artwork can be photocopied and enlarged, or it can be downloaded full-size from www. alabamachanin.com.

Kristina's Rose

To create the projects in this book, Kristina's Rose was enlarged by 472 percent. This artwork can be photocopied and enlarged, or it can be downloaded full-size from www.alabamachanin.com.

Beaded Kristina's Rose Fabric

To create Beaded Kristina's Rose: 1. Stencil fabric using Kristina's Rose stencil (see page 125). **2.** Cut ½" strips of fabric in two colors using instructions for folded stripe appliqué on page 108. **3.** Cut ½" strips of fabric using instructions for stripe with beaded chain stitch on page 105. **4.** Choose one rose shape, and stitch folded -stripe appliqué into place using beaded straight stitch. **5.** Choose another rose, and appliqué stripe with beaded chain stitch. **6.** Choose another rose shape, and sew a beaded rosebud stitch to cover the stenciled shape. **7.** Continue alternating techniques until you have completed all your rose shapes.

Beaded Rosebud Stitch

Folded-Stripe Appliqué with Beaded Straight S...

Stripe Beaded with Chain Stitch

Stars

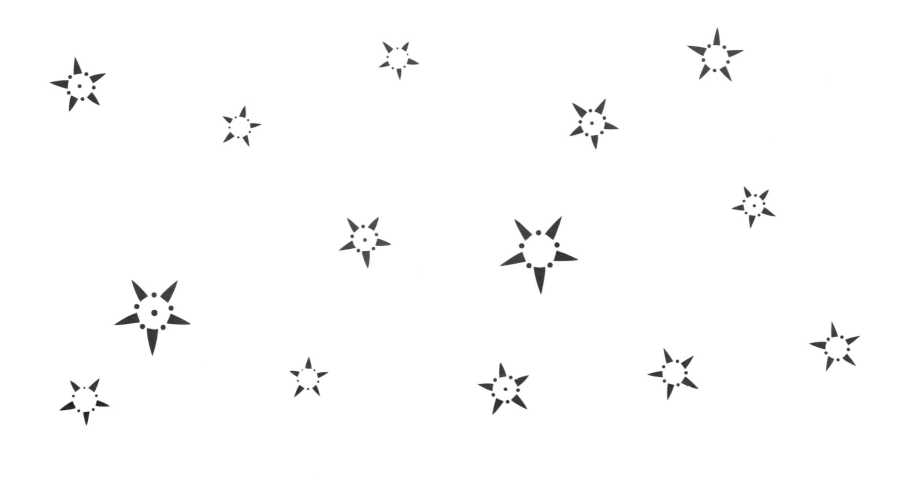

To create the projects in this book, Stars was enlarged by 413 percent. This artwork can be photocopied and enlarged, or it can be downloaded full-size from www.alabamachanin.com.

Satin Stars Fabric

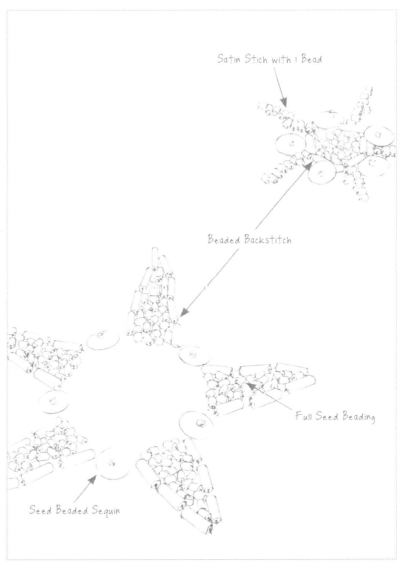

Satin Stitch with 1 Bead

Beaded Backstitch

Full Seed Beading

Seed Beaded Sequin

To create Satin Stars: 1. Stencil fabric using Stars stencil at left. **2.** Fill arms of small star with satin stitch (see page 84), adding one seed or chop bead to each stitch. **3.** Work beaded backstitch (see page 97) with one bugle bead on each stitch around the inner circle of small star. **4.** Fill center circle with full chop or seed beading (see page 73). **5.** For the larger stars, work beaded backstitch around outside edge of each of the arm. **6.** Fill space between beaded backstitch and inner circle with full chop or seed beading. **7.** On each small round dot, add one sequin (see page 76). **8.** Repeat process to cover all star shapes.

To create Climbing Daisy: 1. Stencil fabric using Climbing Daisy from page 132. **2.** Stitch larger petal shapes using cotton tape and a large-eyed embroidery needle.

Stem Stitch

Ribbon Embroidery

French Knot

3. Create French knots (see page 75) with the cotton tape at the center of petal shapes and along stems. **4.** Stem-stitch (see page 85) long, curving stems using embroidery floss. **5.** Repeat process until you have stitched each of your stenciled shapes.

Climbing Daisy

To create the projects in this book, Climbing Daisy was enlarged by 200 percent. This artwork can be photocopied and enlarged, or it can be downloaded full-size from www.alabamachanin.com.

Facets

To create the projects in this book, Facets was enlarged by 475 percent. This artwork can be photocopied and enlarged, or it can be downloaded full-size from www.alabamachanin.com.

Natalie's Dream Fabric

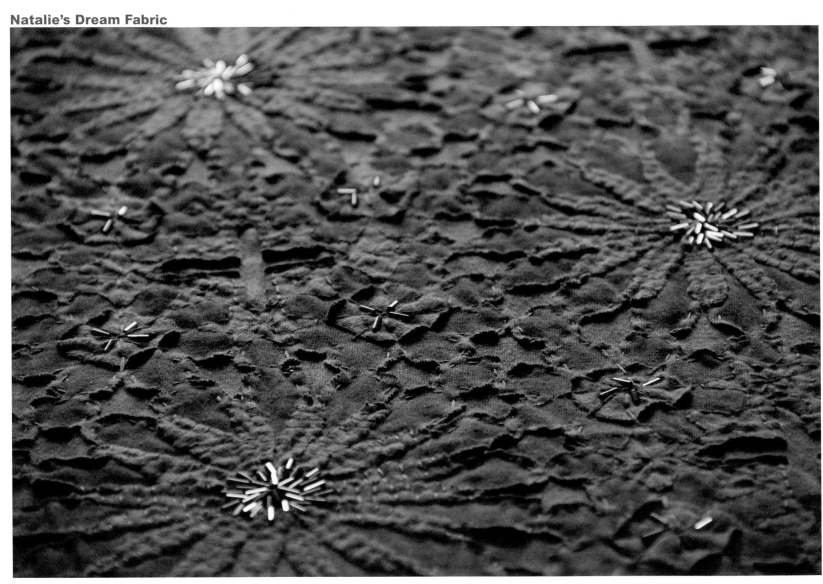

To create Natalie's Dream: 1. Stencil fabric using Facets from page 133. **2.** Negative-reverse-appliqué (see page 99) flower shapes by stitching directly on stenciled line and cutting away shapes between flowers. **3.** Backstitch around cross shapes, between flower shapes, and cut only cross shapes in reverse appliqué inside

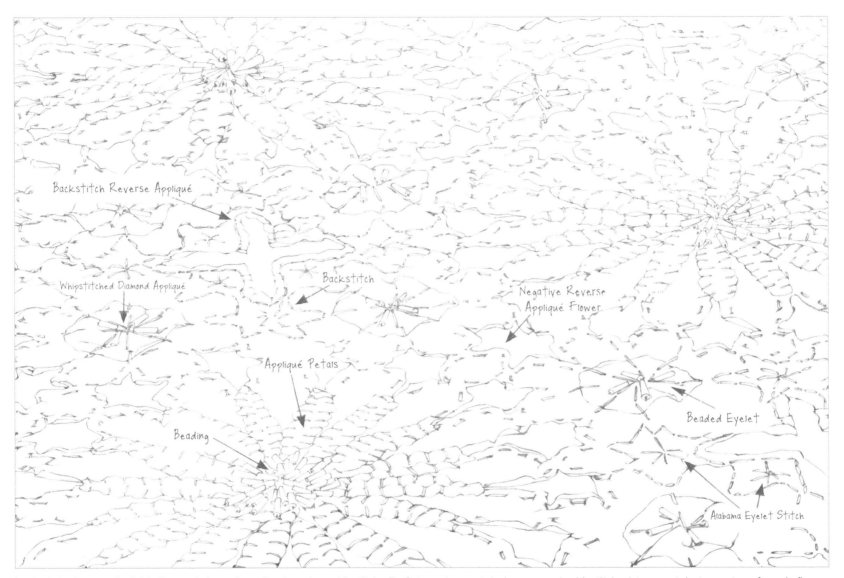

Backstitch Reverse Appliqué

Whipstitched Diamond Appliqué

Backstitch

Negative Reverse Appliqué Flower

Appliqué Petals

Beaded Eyelet

Beading

Alabama Eyelet Stitch

backstitched area. **4.** Add diamond-shaped appliqués using whipstitch. **5.** Cut random petal shapes, and whipstitch sixteen petals to center of each flower. **6.** Add Alabama eyelets (see page 81) where noted. **7.** Stitch beads to center of each flower shape and around the eyelets on each diamond appliqué.

Chapter 9
An Alabama Chanin Wardrobe

The garments we make at Alabama Chanin are simple in shape, flattering, comfortable, and versatile. They are designed to look—and feel—good on a woman's body, whatever her size.

When designing a collection, whether for a fashion season or for this book, I keep in mind my own complicated life. I am a mother, cook, gardener, and chauffeur to my youngest child, as well as an entrepreneur, budding photographer, wannabe filmmaker, and, of course, designer. I want to look beautiful every day but I don't have time to fuss with my clothes, so I need a wardrobe that works for me rather than makes me work. I dare to say that most women have the same requirements.

Through my decade at Alabama Chanin, I have come to love the layering of both embellishments and garments, not only for the way the layering looks, but also for the flexibility it gives a wardrobe. On a day when I don't have a special event to attend, I'm likely to wear basics from Chapter 4 or a lightly embellished piece. I might put on the Fitted Dress with a Bolero or Tied Wrap and boots or sandals. If I have a client meeting, I might take along an embellished Long Skirt like the one in the photo on page 153 and pair it with the basic Ruffle Tunic from page 56. For evening, I choose pieces in subdued colors with more elaborate embroidery and layer them more intricately as in the photos on pages 142–143. And when I'm just running off to the grocery store on a drizzly Saturday morning, I grab a beaded Bucket Hat for me and a basic one for my daughter.

Chapters 1–8 of this book showcase the wide range of techniques we use to create our collections. In this chapter, we present our clothing embellished, layered up, mixed, and matched. While some of the garments may seem elaborate, remember that at the core of each one is basic T-shirt fabric: easy to wear, easy to care for, easy to incorporate into everyday life.

Some of the outfits in this chapter are quite elaborate, depicted as I would showcase them at one of our seasonal presentations for journalists and buyers, as a fashion magazine might style them for photography, or as I might wear them for special occasions. Presenting a collection in this way energizes the designer in me and the little girl in me who loves to play dress-up. However, as you view this collection of garments, remember how easily the individual pieces can be incorporated into everyday life. The Short Skirts can be paired with a simple shirt for the office. The Long Skirts can be dressed up for a wedding or down for a barbecue. The Tied Wrap looks great with jeans. I encourage you to make one piece and then another and another, and before long you'll have built a unique and gorgeous wardrobe customized perfectly to your taste and your lifestyle.

As a fashion designer, I am often asked whom I would like to dress by interviewers who expect me to name a celebrity. The answer is that I want to dress women like me, modern women who may not have perfect bodies or stylists to help them make wardrobe choices but who want to make their way through their busy lives with beauty and grace, who want to sustain valuable traditions and live in beautiful clothing as an accessory to their big and beautiful lives.

Natalie

The Long Baby Doll Dress looks casual on its own, but becomes dressier and more dramatic when coupled with the matching Tied Wrap with random ruffles. At left we show the dress layered with a Short Fitted Skirt with a border of beaded backstitched appliqué. I think it makes a beautiful statement for fashion photography but admit it doesn't make much sense for real life.

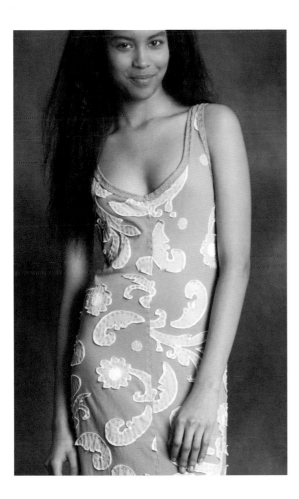

This Long Fitted Dress with a negative reverse appliqué paisley design makes a lovely graphic statement. Wear it with casual flats for day and heels and a matching Poncho for an evening out. I love the subtle interplay between the cream textile paint on the white appliqué fabric and the silt background fabric.

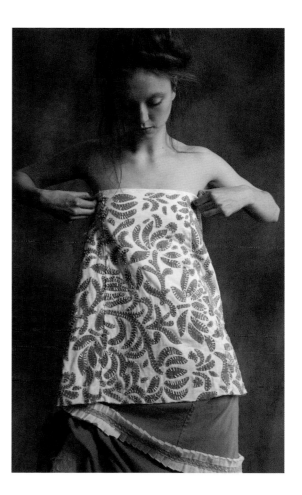

Here we created a variety of evening and bridal looks by layering several different embellished pieces: our Bolero, our Long and Short Skirts (the Short Skirt is worn as a top), and our Fitted Top over a Baby Doll Top. When layering our Long Skirts, I like to shorten the top one a little so that both layers are visible when worn. For these outfits we combined the Anna's Garden and Facets stencils, plus armor beading and beaded appliqué. When combining diverse elements, we use similar colors to create a cohesive effect.

Ribbon embroidery adds beautiful dimension and texture to this Short Skirt. The wide border of beaded backstitched reverse appliqué on the Baby Doll Top is a perfect complement to the fully embellished skirt. I also love to coordinate the Baby Doll Top with a basic Long Skirt like the one on page 162.

The Bucket Hat makes the perfect canvas to test out a variety of fabric designs. It also makes a fantastic gift since it works up quickly and is easily carried in a purse or bag to be worked on during a commute, while talking on the phone, or any other time when your hands are free. The fabric at left is like the one used for the hat at far right except that it is darker and the beads are slightly different.

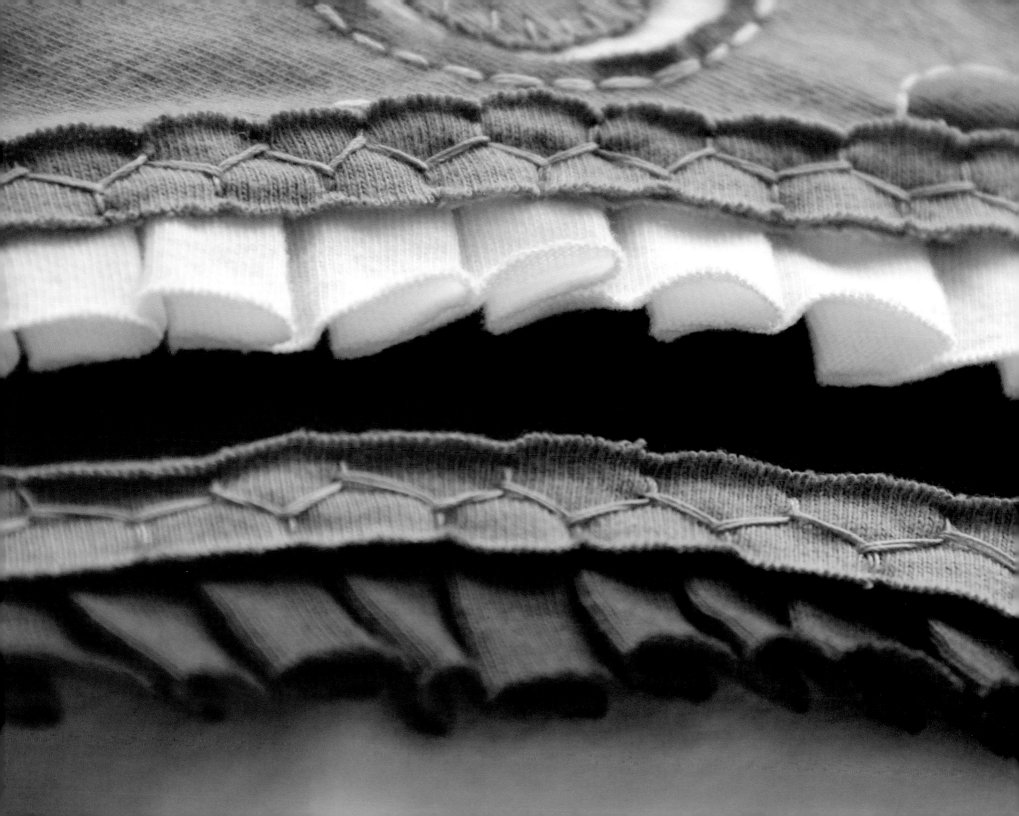

We used the pleated ruffle from page 109 to add a beautiful detail to the neckline of this Fitted Top. This ruffle can be used on the neckline, armhole, or hem of any garment. Simply create a 1"-wide pleated ruffle, align the bottom of it with the bottom of ½"-wide binding, and appliqué the two to the garment using the decorative stitch of your choice. When applying the ruffled ribbing, be sure to overlap the garment edge by ½".

Like the Bucket Hats on pages 146–147, the Short Skirt makes an excellent canvas on which to sample a multitude of fabric designs. The four panels of the skirt work up quickly and are small enough to carry around easily. Here we feature two skirts, one with couching and one with Satin Stars from page 129.

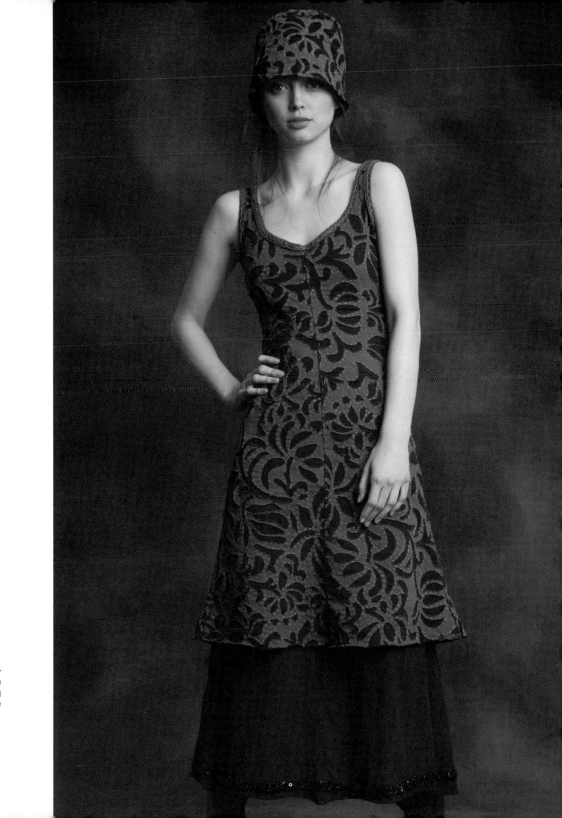

I love the silhouette created by layering a Fitted Dress over our single-layer Mid-Length Skirt. We make the skirt in our lightweight cotton jersey so it has a sliplike quality and add a 1" border of armor beading to give some weight and swing to the hem. Here we chose to pull the outfit together with a Bucket Hat embellished to match the dress.

Like on page 153, here we're layering the Fitted Dress over the Mid-Length Skirt. In this version, we make the skirt more luxurious by outlining each of the stencil shapes with glass beads. I love how the shiny black beads meld together the black and blue fabrics.

We created this very unique look by combining our Tied Wrap, Bolero with long fluted sleeves, Ruffle Top, Short Skirt, and Mid-Length Skirt. The shortness of the Bolero, the Cretan stripes and ruffles on the Ruffle Top, and the beaded border at the bottom of the longer skirt accentuate the long, layered effect. You can also switch up this look with a Tied Wrap in Alabama fur, like the one at far left.

Here our Fitted Top is embroidered with four vertical rows of beaded random ruffles, two on either side of the front seam, and matched with a Short Skirt, Mid-Length Skirt, Tied Wrap, Knotted Necklace, and Fingerless Gloves. All of these pieces can be mixed and matched with the pieces on pages 156 and 157 for even more looks. At right is a close-up of the patterned spiral embroidery on the Short Skirt. This is a beautiful example of how you can create an elegant fabric by carefully combining tones of textile paint, fabric, and thread.

Above the model wears knotted necklaces in a variety of lengths, layered to create a focal point around the neckline. To create the necklaces, cut cotton-jersey ropes (see page 8) from 1" strips of fabric and knot the ropes over and over again to make large knots that look like beads. My favorite necklaces are approximately 36" long and can be wrapped around the neck twice and tied with a bow. You may also want to use these as headbands, bracelets, or belts.

Our Fingerless Gloves can add a touch of edginess to any outfit. These are worked in backstitched reverse applique in a dramatic black on black colorway. See these gloves as part of an entire look on the following page.

We end this chapter with some of my all-time favorite outfits: the Tied Wrap in Natalie's Dream, the Short Fitted Skirt with the beaded ruffle stripe appliqué, and the Fitted Top and Long Skirt worked in negative reverse appliqué using our Anna's Garden stencil. The Long Skirt with the allover Anna's Garden reverse appliqué is the cornerstone of my wardrobe. I have worn it to fancy dinners, on long flights, and even on regular days when I've just wanted to feel comfortable and safe. It is my "uniform." Although the look shown here is quite glamorous, imagine either of these skirts with a white T-shirt or the top with a pair of jeans. It's this type of versatility that I strive for in all of my collections.

Index of Design Choices

At its essence, our design process is a series of choices: garment type, weight and color of fabric, stencil, textile paint color, embroidery techniques, thread and/or floss color, position of knots, seam type. Following are details about every choice we made for the fabrics and garments presented in this book. E-mail us with questions: studio@alabamachanin.com.

Chapter 2 | Stencils & Stenciling

P. 10 Fabric weight–**Medium** | Fabric color–**Silt** | Stencil–**Anna's Garden** | Textile paint–**Pearl Gray** | Thread–**Gray** | Knots–**Inside** | Seams–**Inside Felled** | Binding Stitch–**Cretan P. 15 Fabric 1** Fabric weight–**Medium** | Fabric color–**White** | Stencil–**Facets** | Textile paint–**Cream P. 15 Fabric 2** Fabric weight–**Medium** | Fabric color–**Charcoal** | Stencil–**Facets** | Textile paint–**Cream P. 15 Fabric 3** Fabric weight–**Medium** | Fabric color–**Peacock** | Stencil–**Facets** | Textile paint–**Cream P. 16** Fabric weight–**Medium** | Fabric color–**Steel** | Stencil–**Anna's Garden** | Textile paint–**Pearl Gray P. 18** Fabric weight–**Medium** | Fabric color–**Natural** | Stencil–**Paisley** | Textile paint–**Washed Black** | Marker–**Ultra-Fine Point Black**

P. 18 Fabric weight–**Medium** | Fabric color–**Dark Gray** | Stencil–**Anna's Garden** | Textile paint–**Pearl Gray** | Marker–**Fine Point Black P. 19 Swatch 1** Fabric weight–**Medium** | Fabric color–**Steel** | Stencil–**Anna's Garden** | Textile paint–**Pearl Gray** | Marker–**Fine Point Black P. 19 Swatch 2** Fabric weight–**Medium** | Fabric color–**Silt** | Stencil–**Anna's Garden** | Textile paint–**Pearl Gray** | Marker–**Ultra-Fine Point Black P. 19 Swatch 3** Fabric weight–**Medium** | Fabric color–**Silt** | Stencil–**Anna's Garden** | Textile paint–**Washed Black P. 19 Swatch 4** Fabric weight–**Medium** | Fabric color–**Natural** | Stencil–**Paisley** | Textile paint–**Washed Black** | Marker–**Ultra-Fine Point Black**

Chapter 3 | Basic Stitches

P. 20 Fitted Dress—Fabric weight–**Medium** | Top layer–**Pewter** | Backing–**Pewter** | Stencil–**Spiral + Vitae** | Textile paint–**Pearl Brownie and Slate** | Thread–**Grey** | Embroidery floss–**Mauve** | Knots–**Outside for embroidery, Inside for construction** | Seams–**Inside Felled** | Binding stitch–**Cretan P. 21** Fabric Swatches—Fabric weight–**Medium** | Top layer–**Charcoal** | Backing–**Charcoal** | Stencil–**Anna's Garden** | Textile paint–**Slate** | Thread–**Black** | Knots–**As Noted**

Chapter 4 | Basic Garments & Accessories

P. 30 Model 1 Long Fluted Sleeve Bolero—Fabric weight–**Medium** | Fabric color–**Black** | Thread–**Black** | Knots–**Inside** | Seams–**Inside Felled** | Binding Stitch–**Cretan** | Sleeveless T-Shirt Top—Fabric weight–**Medium** | Fabric color–**Nude** | Thread–**Tan** | Knots–**Inside** | Seams–**Inside Felled** | Binding Stitch–**Cretan** | Mid-Length Skirt—Fabric weight–**Light** | Fabric color–**Deep** | Thread–**Gray** | Knots–**Inside** | Seams–**Inside Felled** | Binding Stitch–**Zigzag** | **Model 2** Cap Sleeve Bolero—Fabric weight–**Medium** | Fabric color–**Sand** | Thread–**Tan** | Knots–**Inside** | Seams–**Inside Felled** | Binding Stitch–**Cretan** | Long Fluted Sleeve T-Shirt Top—Fabric weight–**Medium** | Fabric color–**White** | Thread–**White** | Knots–**Inside** | Seams–**Inside Felled** | Rib–**Cretan** | Mid-Length Skirt—Fabric weight–**Light** | Fabric color–**Black** | Thread–**Black** | Knots–**Inside** | Seams–**Inside Felled** | Binding Stitch–**Zigzag P. 48 Model 1** Long Fluted Sleeve T-Shirt Top—Fabric weight–**Medium** | Fabric color–**White** | Thread–**White** | Knots–**Inside** | Seams–**Inside Felled** | Binding stitch–**Cretan** | **Model 2** Sleeveless T-Shirt Top—Fabric weight–**Medium** | Fabric color–**Nude** | Thread–**Tan** | Knots–**Inside** | Seams–**Inside Felled** | Binding stitch–**Cretan P. 52** Long Fitted Dress—Fabric weight–**Medium** | Fabric color–**Navy** | Thread–**Navy** | Knots–**Inside** | Seams–**Inside Felled** | Binding stitch–**Cretan P. 53** Short Fitted Dress—Fabric weight–**Medium** | Fabric color–**Ochre** | Thread–**Tan** | Knots–**Inside** | Seams–**Inside Felled** | Binding Stitch–**Cretan P. 54** Fitted Tunic—Fabric weight–**Medium** | Fabric color–**Deep** | Thread–**Gray** | Knots–**Inside** | Seams–**Inside Felled** | Binding Stitch–**Cretan P. 55** Fitted Tunic—Fabric weight–**Medium** | Fabric color–**Deep** | Thread–**Gray** | Knots–**Inside** | Seams–**Inside Felled** | Binding Stitch–**Cretan**

P. 55 Fitted Top—Fabric weight–**Medium** | Fabric color–**Sand** | Thread–**Tan** | Knots–**Inside** | Seams–**Inside Felled** | Binding Stitch–**Cretan** | Mid-Length Skirt—Fabric weight–**Light** | Fabric color–**Black** | Thread–**Black** | Knots–**Inside** | Seams–**Inside Felled** | Binding Stitch–**Zigzag P. 56** Fitted Baby Doll Top —Fabric weight–**Medium** | Fabric color–**Deep** | Thread–**Gray** | Knots–**Inside** | Seams–**Inside Felled** | Binding stitch–**Cretan** | Ruffle stitch–**Zigzag Chain** | Mid-Length Skirt—Fabric weight–**Light** | Fabric color–**Black** | Thread–**Black** | Knots–**Inside** | Seams–**Inside Felled** | Binding stitch–**Zigzag P. 56** Fitted Top w/ Ruffle—Fabric weight–**Medium** | Fabric color–**Nude** | Thread–**Tan** | Knots–**Inside** | Seams–**Inside Felled** | Binding stitch–**Cretan** | Ruffle stitch–**Zigzag Chain** | Mid-Length Skirt—Fabric weight–**Light** | Fabric color–**Deep** | Thread–**Gray** | Knots–**Inside** | Seams–**Inside Felled** | Binding stitch–**Zigzag P. 58** Fitted Baby Doll Dress—Fabric weight–**Medium** | Fabric color–**Black** | Thread–**Black** | Knots–**Inside** | Seams–**Inside Felled** | Binding stitch–**Cretan** | Ruffle stitch–**Zigzag Chain P. 58** Long Fitted Baby Doll Dress—Fabric weight–**Medium** | Fabric color–**Nude** | Thread–**Pink** | Knots–**Inside** | Seams–**Inside Felled** | Binding stitch–**Cretan** | Ruffle stitch–**Zigzag Chain P. 59** Peasant Dress—Fabric weight–**Medium** | Fabric color–**Black** | Third contrast ruffle–**Deep** | Thread–**Black and Gray** | Knots–**Inside** | Seams–**Inside Felled** | Binding stitch–**Cretan** | Ruffle stitch–**Zigzag Chain**

P. 60 Short Skirt—Fabric weight–**Medium** | Fabric color–**Carmine** | Thread–**Red** | Knots–**Inside** | Seams–**Inside Felled** | Binding Stitch–**Zigzag** | Cap Sleeve T-Shirt Top—Fabric weight–**Medium** | Fabric color–**Deep** | Thread–**Gray** | Knots–**Inside** | Seams–**Inside Felled** | Binding Stitch–**Cretan P. 61** Mid-Length Skirt—Fabric weight–**Medium** | Fabric color–**Silt** | Thread–**Gray** | Knots–**Inside** | Seams–**Inside Felled** | Binding Stitch–**Zigzag** | Long Fluted Sleeve T-Shirt Top—Fabric weight–**Medium** | Fabric color–**White** | Thread–**White** | Knots–**Inside** | Seams–**Inside Felled** | Binding Stitch–**Cretan P. 62** Long Skirt—Fabric weight–**Medium** | Fabric color–**Indigo** | Thread–**Gray** | Knots–**Inside** | Seams–**Inside Felled** | Binding Stitch–**Zigzag** | Sleeveless T-Shirt Top—Fabric weight–**Medium** | Fabric color–**Nude** | Thread–**Tan** | Knots–**Inside** | Seams–**Inside Felled** | Binding Stitch–**Cretan P. 63** Long Fluted Sleeve Bolero—Fabric weight–**Light** | Fabric color–**Black** | Thread–**Black** | Chop Beads–**Black** | Bugle Beads–**#3 Black** | Sequins–**Black & Dark Brown** | Knots–**Inside** | Seams–**Inside Felled** | Binding stitch–**Cretan**

Mid-Length Skirt—Fabric weight–**Light** | Fabric color–**Black** | Thread–**Black** | Knots–**Inside** | Seams–**Inside Felled** | Binding stitch–**Zigzag P. 64** Poncho—Fabric weight–**Medium** | Fabric color–**Black** | Thread–**Black** | Knots–**Inside** | Seams–**Outside Floating** | Fitted Tunic—Fabric weight–**Medium** | Fabric color–**Deep** | Thread–**Gray** | Knots–**Inside** | Seams–**Inside Felled** | Binding Stitch–**Cretan**

P. 65 Poncho—Fabric weight–**Medium** | Fabric color–**Black** | Thread–**Black** | Knots–**Inside** | Seams–**Outside Floating** | Fitted Top—Fabric weight–**Medium** | Fabric color–**Deep** | Thread–**Gray** | Knots–**Inside** | Seams–**Inside Felled** | Binding Stitch–**Cretan** | Mid-Length Skirt—Fabric weight–**Light** | Fabric color–**Black** | Thread–**Black** | Knots–**Inside** | Seams–**Inside Felled** | Binding Stitch–**Zigzag P. 66 & 67** Tied Wrap—Fabric weight–**Medium** | Top-layer–**Black** | Backing–**Black** | Inside–**Carmine** | Thread–**Black** | Knots–**Inside** | Seams–**Outside Floating** | Baby Doll Top—Fabric weight–**Medium** | Fabric color–**Deep** | Thread–**Gray** | Knots–**Inside** | Seams–**Inside Felled** | Binding Stitch–**Cretan** | Ruffle Stitch–**Zigzag Chain** | Mid-Length Skirt—Fabric weight–**Light** | Fabric color–**Black** | Thread–**Black** | Knots–**Inside** | Seams–**Inside Felled** | Binding Stitch–**Zigzag P. 68** Bucket Hat—Fabric weight–**Medium** | Top-layer–**Black** | Backing–**Black** | Contrasting band–**Navy** | Thread–**Black for construction, Navy for contrasting band** | Knots–**Inside** | Seams–**Inside Floating** | Binding Stitch–**Herringbone P. 70** Fingerless Gloves—Fabric weight–**Medium** | Top-layer–**Silt** | Bottom-layer–**Silt** | Thread–**Gray** | Knots–**Inside** | Seams–**Inside Felled** | Fitted Dress—Fabric weight–**Medium** | Fabric color–**Black** | Thread–**Black** | Knots–**Inside** | Seams–**Inside Felled** | Binding stitch–**Cretan**

Chapter 5 | Embellishing with Beads, Sequins & Embroidery

P. 72 Fabric weight–**Medium** | Top layer–**Dove** | Backing–**Dove** | Stencil–**Anna's Garden** | Textile paint–**Pearl Gray** | Thread–**Gray** | Knots–**Inside** | Bugle bead–**#3 Satin Gray** | Chop bead–**Clear Black** | Sequin–**Medium Silver P. 73 & P. 74 Beading Swatches** Fabric weight–**Medium** | Top layer–**Sand** | Backing–**Sand** | Thread–**Gray** | Knots–**Inside** | Bugle bead–**#3 Satin Gray** | Chop bead–**Clear Black P. 75** Fabric weight–**Medium** | Top layer–**Charcoal** | Backing–**Charcoal** | Stencil–**Abbie's Flower** | Textile paint–**Pearl Brownie** | Thread–**Black** | Knots–**Inside** | French knots–**Outside P. 76 & P. 77 Sequin Swatches** Fabric weight–**Medium** | Top layer–**Sand** | Backing–**Sand** | Thread–**Gray** | Knots–**Inside** | Bugle bead–**#3 Satin Gray** | Chop bead–**Clear Black** | Sequin–**Medium Silver P. 78 & P. 79 Stitched Swatches** Fabric Weight–**Medium** | Top layer–**White** | Backing–**White** | Thread–**Brown** | Knots–**Inside** | Chop beads–**Brown**

P. 80 Fabric weight–**Medium** | Top-layer–**Deep** | Backing–**Deep** | Thread–**Black** | Embroidery floss–**Black, Dark Gray, Medium Gray, Tea** | Knots–**Inside** | Bugle bead–**#3 Brown P. 82** Fabric weight–**Medium** | Top-layer–**Black** | Backing–**Black** | Stencil–**Spirals** | Textile paint–**Slate** | Embroidery floss–**Variegated Black** | Knots–**Outside P. 83** Fabric weight–**Medium** | Top-layer–**Black** | Backing–**Black** | Stencil–**Spirals** | Textile paint–**Pearl Brownie** | Embroidery Floss–**Brown** | Knots–**Inside P. 84** Fabric weight–**Medium** | Top-layer–**Sand** | Backing–**Sand** | Stencil–**Abbie's Flower** | Textile paint–**Pearl Silver** | Thread–**Cream** | Knots–**Inside P. 85** Fabric weight–**Medium** | Top-layer–**Ochre** | Backing–**Ochre** | Stencil–**Abbie's Flower** | Textile paint–**Charcoal** | Embroidery Floss–**Dark Gray** | Knots–**Inside P. 88** Fabric weight–**Medium** | Top-layer–**Earth** | Backing–**Earth** | Stencil–**Spirals and Vitae** | Textile paint–**Pearl Brownie and Slate** | Embroidery floss–**Black** | Knots–**Outside**

Chapter 6 | Quilting & Reverse Appliqué

P. 92 Fitted Dress—Fabric weight–**Medium** | Top-layer–**Sand** | Backing–**White** | Binding–**White** | Stencil–**Anna's Garden** | Textile paint–**Pearl Silver** | Thread–**Tan & White** | Knots–**Inside** | Seams–**Inside Felled** | Binding stitch–**Cretan P. 93** Fabric weight–**Medium** | Top-layer–**Earth** | Backing–**Earth** | Stencil–**Anna's Garden** | Textile paint–**Pearl Brownie** | Thread–**Brown** | Knots–**Inside P. 94** Fabric weight–**Medium** | Top-layer–**Ruby** | Backing–**Ruby** | Stencil–**Anna's Garden** | Textile paint–**Burgundy** | Thread–**Burgundy** | Knots–**Inside**

P. 96 Swatch 1 Fabric weight–**Medium** | Top-layer–**Silt** | Backing–**Black** | Stencil–**Anna's Garden** | Textile paint–**Charcoal** | Thread–**Black** | Knots–**Inside** | Bugle beads–**#3 Black P. 96 Swatch 2** Fabric weight–**Medium** | Top-layer–**Navy** | Backing–**Black** | Stencil–**Anna's Garden** | Textile paint–**Black** | Thread–**Black** | Knots–**Inside** | Chop beads–**Black**

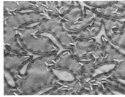

P. 96 Swatch 3 Fabric weight–**Medium** | Top-layer–**Sand** | Backing–**White** | Stencil–**Anna's Garden** | Textile paint–**Pearl Gray** | Thread–**Tan** | Knots–**Inside** | Bugle beads–**#3 Satin White P. 96 Swatch 4** Fabric weight–**Medium** | Top-layer–**Black** | Backing–**Black** | Stencil–**Anna's Garden** | Textile paint–**Pearl Brownie** | Thread–**Black** | Knots–**Inside** | Bugle beads–**#3 Brown P. 97 Swatch 1** Fabric weight–**Medium** | Top-layer–**Forest** | Backing–**Forest** | Stencil–**Anna's Garden** | Textile paint–**Pearl Brownie** | Thread–**Brown** | Knots–**Inside** | Bugle beads–**#3 Brown**

P. 97 Swatch 2 Fabric weight–**Medium** | Top-layer–**White** | Backing–**Sand** | Stencil–**Anna's Garden** | Textile paint–**Putty** | Thread–**White** | Embroidery floss–**Cream** | Knots–**Inside P. 97 Swatch 3** Fabric weight–**Medium** | Top-layer–**Doeskin** | Backing–**White** | Stencil–**Anna's Garden** | Textile paint–**Putty** | Embroidery floss–**Tea** | Knots–**Inside P. 97 Swatch 4** Fabric weight–**Medium** | Top-layer–**Silt** | Backing–**Black** | Stencil–**Anna's Garden** | Textile paint–**Slate** | Thread–**Black** | Knots–**Inside** | Bugle beads–**#3 Black**

P. 98 Swatch 1 Fabric weight–**Medium** | Top-layer–**Black** | Backing–**Earth** | Stencil–**Anna's Garden** | Textile paint–**Pearl Brownie** | Thread–**Black** | Knots–**Inside** | Seed beads–**Brown P. 98 Swatch 2** Fabric weight–**Medium** | Top-layer–**White** | Backing–**Twilight** | Stencil–**Anna's Garden** | Textile paint–**Pearl Gray** | Thread–**White** | Knots–**Inside** | Chop beads–**White** | Marker–**Ultra-Fine Point Brown P. 98 Swatch 3** Fabric weight–**Medium** | Top-layer–**Twilight** | Backing–**Twilight** | Stencil–**Anna's Garden** | Textile paint–**Black** | Thread–**Black** | Knots–**Inside** | Bugle beads–**#3 Black** | Chop beads–**Black** | Marker–**Ultra-Fine Point Black**

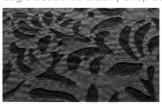  

P. 98 Swatch 4 Fabric weight–**Medium** | Top-layer–**Twilight** | Backing–**Black** | Stencil–**Anna's Garden** | Textile paint–**Slate** | Thread–**Gray** | Knots–**Inside P. 99** Fabric weight–**Medium** | Top-layer–**Midnight** | Backing–**Midnight** | Stencil–**Anna's Garden** | Textile paint–**Pearl Gray** | Thread–**Gray** | Knots–**Outside P. 99 Poncho**—Fabric weight–**Medium** | Top-layer–**Forest** | Backing–**Black** | Stencil–**Anna's Garden** | Textile paint–**Pearl Brownie** | Thread–**Black** | Knots–**Outside**

Chapter 7 | Appliqué

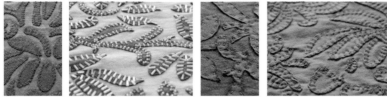

P. 100 Fabric weight–**Medium** | Top-layer–**Silt** | Backing–**Silt** | Appliqué–**Silt** | Stencil–**Anna's Garden** | Textile paint–**Pearl Gray** | Thread–**Gray** | Knots–**Inside** **P. 101 Swatch 1** Fabric weight–**Medium** | Top-layer–**White** | Backing–**White** | Appliqué–**Silt** | Stencil–**Anna's Garden** | Textile paint–**Putty** | Thread–**White** | Knots–**Inside** | Bugle beads–**#3 Satin White P. 101 Swatch 2** Fabric weight–**Medium** | Top-layer–**Deep** | Backing–**Deep** | Appliqué–**Deep** | Stencil–**Anna's Garden** | Textile paint–**Pearl Gray** | Thread–**Gray** | Knots–**Inside P. 102 Swatch 1** Fabric weight–**Medium** | Top-layer–**Silt** | Backing–**Silt** | Appliqué–**Silt** | Stencil–**Anna's Garden** | Textile paint–**Pearl Gray** | Thread–**Gray** | Knots–**Inside**

P. 102 Swatch 2 Fabric weight–**Medium** | Top-layer–**Silt** | Backing–**Silt** | Appliqué–**Silt** | Stencil–**Anna's Garden** | Textile paint–**Pearl Gray** | Thread–**Gray** | Knots–**Inside** | Chop beads–**Clear Black P. 102 Swatch 3** Fabric weight–**Medium** | Top-layer–**Black** | Backing–**Black** | Appliqué–**Charcoal** | Stencil–**Anna's Garden** | Textile paint–**Pearl Gray** | Embroidery Floss–**Black** | Knots–**Inside P. 102 Swatch 4** Fabric weight–**Medium** | Top-layer–**Nude** | Backing–**Nude** | Appliqué–**Nude** | Stencil–**Anna's Garden** | Textile paint–**Pearl Gray** | Thread–**Tan** | Knots–**Inside** | Chop beads–**Pink P. 103** Fabric weight–**Medium** | Top-layer–**Silt** | Backing–**Silt** | Appliqué–**Black** | Stencil–**Anna's Garden** | Textile paint–**Slate** | Thread–**Black** | Knots–**Inside**

P. 103 Fabric weight–**Medium** | Top-layer–**Black** | Backing–**Black** | Appliqué–**Black** | Stencil–**Anna's Garden** | Textile paint–**Slate** | Thread–**Black** | Knots–**Inside** | Bugle beads–**#3 Black** | Chop beads–**Black P. 104** Fabric weight–**Medium** | Top-layer–**Doeskin** | Backing–**Doeskin** | Appliqué–**Midnight** | Thread–**Gray** | Knots–**Inside** | Stitch–**Cretan**

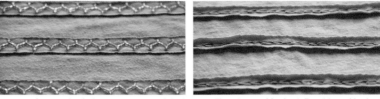

P. 105 Swatch 1 Fabric weight–**Medium** | Top-layer–**Nude** | Backing–**Nude** | Appliqué–**Nude** | Thread–**Tan** | Knots–**Inside** | Stitch–**Cretan** | Chop beads–**Pink P. 105 Swatch 2** Fabric weight–**Medium** | Top-layer–**Sand** | Backing–**Sand** | Appliqué–**Sand** | Thread–**Tan** | Knots–**Inside** | Stitch–**Chain**

P. 105 Swatch 3 Fabric weight–**Medium** | Top-layer–**Black** | Backing–**Black** | Appliqué–**Black** | Thread–**Black** | Knots–**Inside** | Stitch–**Chain** | Chop beads–**Black P. 106** Fabric weight–**Medium** | Top-layer–**Nude** | Backing–**Nude** | Appliqué–**Nude** | Thread–**Tan** | Knots–**Inside** | Stitch–**Straight**

P. 107 Fabric weight–**Medium** | Top-layer–**Black** | Backing–**Black** | Appliqué–**Black** | Thread–**Black** | Knots–**Inside** | Stitch–**Straight** | Bugle beads–**#3 Black** | Chop beads–**Black P. 108 Swatch 1** Fabric weight–**Medium** | Top-layer–**White** | Backing–**White** | Appliqué #1–**Silt** | Appliqué #2–**Sand** | Thread–**Tan** | Knots–**Inside**

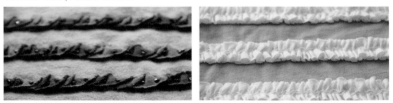

P. 108 Swatch 2 Fabric weight–**Medium** | Top-layer–**Charcoal** | Backing–**Charcoal** | Appliqué #1–**Black** | Appliqué #2–**Charcoal** | Thread–**Black** | Knots–**Inside** | Chop beads–**Black P. 108 Swatch 3** Fabric weight–**Medium** | Top-layer–**Sand** | Backing–**Sand** | Appliqué–**White** | Thread–**White** | Knots–**Inside** | Stitch–**Zigzag Chain**

P. 108 **Swatch 4** Fabric weight–**Medium** | Top-layer–**Nude** | Backing–**Nude** | Appliqué–**Nude** | Thread–**Tan** | Knots–**Inside** | Stitch–**Zigzag Chain** | Chop beads–**Pink** P. 109 Fabric weight–**Medium** | Top-layer–**Black** | Backing–**Black** | Appliqué #1–**Midnight** | Appliqué #2–**Sand** | Thread–**Tan** | Stitch–**Cretan** | Knots–**Inside**

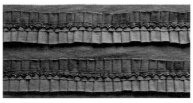

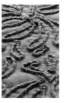

P. 109 Fabric weight–**Medium** | Top-layer–**Black** | Backing–**Black** | Appliqué #1–**Black** | Appliqué #2–**Black** | Thread–**Black** | Knots–**Inside** | Stitch–**Cretan** | Chop beads–**Black** P. 110 Fabric weight–**Medium** | Top-layer–**Midnight** | Backing–**Midnight** | Couching ropes–**Midnight** | Stencil–**Anna's Garden** | Textile paint–**Pearl Gray** | Thread–**Gray** | Knots–**Inside** P. 111 **Swatch 1** Fabric weight–**Medium** | Top-layer–**Nude** | Backing–**Nude** | Couching ropes–**Nude** | Stencil–**Anna's Garden** | Textile paint–**Pearl Gray** | Thread–**Tan** | Knots–**Inside** | Chop beads–**Pink** P. 111 **Swatch 2** Fabric weight–**Medium** | Top-layer–**Silt** | Backing–**Silt** | Couching ropes–**Black Cotton Yarn** | Stencil–**Anna's Garden** | Textile paint–**Slate** | Thread–**Black** | Knots–**Inside**

P. 113 Fabric weight–**Medium** | Top-layer–**Dove** | Backing–**Dove** | Appliqué–**Dove** | Thread–**Gray** | Knots–**Inside** | Stitch–**Herringbone** P. 114 Fabric weight–**Medium** | Top-layer–**Black** | Backing–**Black** | Appliqué #1–**Black** | Appliqué #2–**Silt** | Thread–**Black** | Stitch–**Straight** | Knots–**Inside** P. 115 Ruffle Passementerie Scarf—Fabric weight–**Medium** | Top-layer–**Plum** | Bottom-layer–**Plum** | Appliqué #1–**Plum** | Appliqué #2–**Deep** | Thread–**Burgundy** | Knots–**Inside** | Seams–**Outside Floating**

Chapter 8 | Fabrics & Fabric Maps

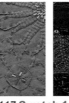

P. 117 **Swatch 1** Fabric weight–**Medium** | Top-layer–**Silt** | Backing–**Silt** | Appliqué–**Silt** | Stencil–**Facets** | Textile paint–**Pearl Gray** | Thread–**Gray** | Embroidery floss–**Lt. Gray** | Knots–**Inside** | Bugle beads–**#3 Satin Gray** P. 117 **Swatch 2** Fabric weight **Medium** | Top-layer–**Twilight** | Backing–**Black** | Appliqué #1–**Twilight** | Appliqué #2–**Black** | Stencil–**June's Spring** | Textile paint–**Charcoal** | Thread–**Black and Gray** | Embroidery floss–**Lt. Gray** | Knots–**Inside** | Bugle beads–**#3 Black** | Chop beads–**Black** | Sequins–**Black and Dk. Brown** P. 117 **Swatch 3** Fabric weight–**Medium** | Top-layer–**White** | Backing–**White** | Appliqué #1–**Black** | Appliqué #2–**Charcoal** | Appliqué #3–**Deep** | Stencil–**Kristina's Rose** | Textile paint–**Pearl Silver** | Thread–**Black** | Knots–**Inside** | Chop beads–**Black** P. 117 **Swatch 4** Fabric weight–**Medium** | Top-layer–**Silt** | Backing–**Silt** | Stencil–**Climbing Daisy** | Textile paint–**Pearl Gray** | Embroidery floss–**Lt. Gray** | Knots–**Inside** | Ribbon–**Gray** P. 118 Fabric weight–**Medium** | Top-layer–**Black** | Backing–**Deep** | Appliqué #1–**Deep** | Appliqué #2–**Black** | Stencil–**June's Spring** | Textile paint–**Slate** | Thread–**Black and Gray** | Embroidery floss–**Black** | Knots–**Inside** | Bugle beads–**#3 Black** | Chop beads–**Black** |Sequins–**Black and Dk. Brown** P. 121 Fabric weight–**Medium** |Top-layer–**Silt** | Backing–**Silt** | Stencil–**Fern** | Textile paint–**Pearl Gray** | Thread–**Gray** | Knots–**Outside** | Chop beads–**Silver**

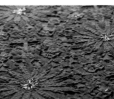

P. 122 Fabric weight–**Medium** | Top-layer–**Silt** | Backing–**White** | Stencil–**Paisley** | Textile paint–**Cream** | Embroidery floss–**Silt** | Thread–**Tan** | Knots–**Inside** | Chop beads–**Silver** | Bugle beads–**#2 Silver** | Sequins–**Iridescent** P. 126 Fabric weight–**Medium** | Top-layer–**Black** | Backing–**Black** | Appliqué #1–**Black** | Appliqué #2–**Charcoal** | Appliqué #3–**Deep** | Stencil–**Kristina's Rose** | Textile paint–**Slate** | Thread–**Black** | Knots–**Inside** | Chop beads–**Black** P. 129 Fabric weight–**Medium** | Top-layer–**Black** | Backing–**Black** | Stencil–**Stars** | Textile paint–**Pearl Brownie** | Thread–**Brown** | Knots–**Inside** | Bugle beads–**#3 Brown** | Chop beads–**Brown** | Sequins–**Iridescent** P. 130 Fabric weight–**Medium** | Top-layer–**Black** | Backing–**Black** | Stencil–**Climbing Daisy** | Textile paint–**Slate** | Embroidery floss–**Lt. Gray** | Knots–**Inside** | Ribbon–**Gray** P. 134 Fabric weight–**Light** | Top-layer–**Black** | Backing–**Black** | Appliqué–**Black** | Stencil–**Facets** | Textile paint–**Slate** | Thread–**Black** | Embroidery floss–**Black** | Knots–**Inside** | Bugle beads–**#3 Black**

Chapter 9 | An Alabama Chanin Wardrobe

P. 138 & P. 139 Tied Wrap—Fabric weight—**Medium** | Top-layer—**Nude** | Backing—**Nude** | Appliqué—**Nude** | Treatment—**Beaded Random Ruffle** | Stitch—**Straight** | Thread—**Tan** | Seams—**Outside Floating** | Chop beads—**Pink** | Long Fitted Baby Doll Dress—Fabric weight—**Medium** | Fabric color—**Nude** | Thread—**Tan** | Seams—**Inside Felled** | Binding Stitch—**Cretan** | Ruffle Stitch—**Zigzag Chain** | Short Skirt—Fabric Weight—**Medium** | Top-layer—**Nude** | Backing—**Nude** | Appliqué—**Nude** | Stencil—**Anna's Garden** | Treatment—**Appliqué with Beaded Backstitch** | Textile paint—**Pearl Silver** | Thread—**Tan** | Seams—**Inside Felled** | Binding Stitch—**Zigzag** | Chop beads—**Pink**

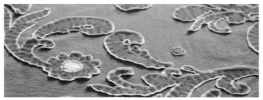

P. 140 Fabric weight—**Medium** | Top-layer—**White** | Backing—**Silt** | Appliqué—**White** | Stencil—**Paisley** | Treatment—**Negative Reverse Appliqué** | Textile Paint—**Cream** | Thread—**Tan** | Knots—**Inside P. 141** Long Fitted Dress | Fabric weight **Medium**— Top-layer—**White** | Backing—**Silt** | Appliqué—**White** | Stencil—**Paisley** | Treatment—**Negative Reverse Appliqué** | Textile paint—**Cream** | Thread—**Tan** | Knots—**Inside** | Seams—**Inside Felled** | Binding Stitch—**Cretan**

P. 142 & P. 143 Model 1 Sleeveless Bolero—Fabric weight—**Medium** | Top-layer—**White** | Backing—**White** | Stencil—**Spiral** | Textile paint—**Putty** | Treatment—**Embroidery** | Thread—**White** | Embroidery floss—**Cream** | Knots—**Outside** | Seams—**Inside Felled** | Binding Stitch—**Cretan** | Short Skirt—Fabric weight—**Medium** | Top-layer—**White** | Backing—**White** | Appliqué—**Silt** | Stencil—**Anna's Garden** | Treatment—**Beaded Appliqué** | Thread—**White** | Knots—**Inside** | Seams—**Inside Felled** | Binding Stitch—**Zigzag** | Bugle beads—**#3 White** | Chop Beads—**White** | Long Skirt—Fabric weight—**Medium** | Fabric color—**Silt** | Appliqué #1—

White | Appliqué #2—**Sand** | Treatment—**Beaded Pleated Ruffle** | Stitch—**Cretan** | Thread—**Tan** | Seams—**Inside Felled** | Binding Stitch—**Zigzag** | Chop beads—**White** | Long Skirt—Fabric weight—**Medium** | Top-layer—**White** | Backing—**White** | Stencil—**Facets** | Treatment—**Negative Reverse Appliqué** | Textile paint—**Putty** | Thread—**White** | Knots—**Outside** | Seams—**Inside Felled** | Binding Stitch—**Zigzag** | **Model 2** Fitted Top—Fabric weight—**Medium** | Fabric color—**Silt** | Treatment—**Fade Beaded** | Thread—**Tan** | Knots—**Inside** | Seams—**Inside Felled** | Binding Stitch—**Beaded Cretan** | Chop beads—**White** | Bugle beads—**#3 White** | Sequins—**Clear and iridescent** | Baby Doll Top—Fabric weight—**Medium** | Fabric color—**Silt** | Thread—**Tan** | Knots—**Inside** | Seams—**Inside Felled** | Binding Stitch—**Cretan** | Ruffle Stitch—**Zigzag Chain** | Long Skirt—Fabric weight—**Medium** | Top-layer—**Silt** | Backing—**Pewter** | Stencil—**Anna's Garden** | Treatment—**Inked & Beaded Backstitch Reverse Appliqué with Inside Beading** | Textile paint—**Pearl Medium Brown** | Thread—**Tan and White** | Knots—**Inside** | Seams—**Inside Felled** | Binding Stitch—**Zigzag** | Chop beads—**White** | Bugle beads **#3 White** | Marker—**Ultra-Fine Point Brown** | Fabric swatch—Fabric weight—**Medium** | Top-layer—**Silt** | Backing—**Pewter** | Stencil—**Anna's Garden** | Treatment—**Inked & Beaded Backstitch Reverse Appliqué with Inside Beading** | Textile Paint—**Pearl Medium Brown** | Thread—**Tan and White** | Knots—**Inside** | Chop beads—**White** | Bugle beads—**#3 White** | Marker—**Ultra-Fine Point Brown**

P. 144 Fabric weight—**Medium** | Top-layer—**Silt** | Backing—**Silt** | Stencil—**Climbing Daisy** | Treatment—**Ribbon Embroidery** | Textile paint—**Pearl Grey** | Embroidery floss—**Lt. Grey** | Knots—**Inside** | Ribbon—**Grey P. 145** Fitted Baby Doll Top—Fabric weight—**Medium** | Top-layer—**Deep** | Appliqué—**Deep** | Stencil—**Anna's Garden** | Treatment—**Appliqué with Beaded Straight Stitch** | Textile paint—**Pearl Charcoal** | Thread—**Grey** | Knots—**Inside** | Seams—**Inside Felled** | Binding Stitch—**Cretan** | Ruffle Stitch—**Zigzag Chain** | Chop Beads—**Clear Black** | Short Skirt—Fabric weight—**Medium** | Top-layer—**Sand** | Backing—**Sand** | Stencil—**Climbing Daisy** | Treatment—**Ribbon Embroidery** | Textile paint—**Pearl Grey** | Thread—**Tan** | Embroidery floss—**Tea** | Ribbon—**Cream** | Knots—**Inside** | Seams—**Inside Felled** | Binding Stitch—**Zigzag**

P. 146 Fabric Weight—**Medium** | Top-layer—**Black** | Backing—**Black** | Stencil—**Anna's Garden** | Treatment—**Accent-Beaded Reverse Appliqué** | Textile paint—**Pearl**

Brownie | Thread–**Black** | Knots–**Inside** | Bugle Beads–**#3 Brown** **P. 146 Model 1** Fitted Top—Fabric weight–**Medium** | Fabric color–**Deep** | Thread–**Grey** | Knots– **Inside** | Seams–**Inside Felled** | Binding Stitch–**Cretan** | Bucket Hat—Fabric weight–**Medium** | Top-layer–**Deep** | Backing–**Deep** | Stencil–**Spiral** | Treatment– **Embroidery** | Textile paint–**Slate** | Thread–**Black** | Embroidery floss–**Variegated** | Knots–**Outside** | Seams–**Inside Floating** | **Model 2** Fitted Top—Fabric weight– **Medium** | Fabric color–**Black** | Thread–**Black** | Knots–**Inside** | Seams–**Inside Felled** | Binding Stitch–**Cretan** | Bucket Hat—Fabric weight–**Medium** | Top-layer– **Black** | Backing–**Black** | Appliqué–**Black** | Stencil–**June's Spring** | Treatment– **Mixed** | Textile paint–**Slate** | Thread–**Black** | Embroidery floss–**Black** | Knots– **Inside** | Seams–**Inside Floating** | Chop Beads–**Black** | Bugle Beads–**#3 Black** | Sequins–**Black and Dk. Brown**

P. 147 Model 1 Bucket Hat—Fabric weight–**Medium** | Top-layer–**Deep** | Backing– **Deep** | Stencil–**Fern** | Treatment–**Embroidery** | Textile paint–**Pearl Charcoal** | Thread–**Black** | Knots–**Inside** | Seams–**Inside Floating** | Fitted Top—Fabric weight–**Medium** | Fabric color–**Deep** | Thread–**Grey** | Knots–**Inside** | Seams– **Inside Felled** | Binding Stitch–**Cretan** | **Model 2** Bucket Hat—Fabric weight– **Medium** | Top-layer–**Deep** | Backing–**Deep** | Stencil–**Anna's Garden** | Treatment– **Accent Beaded Reverse Appliqué** | Textile paint–**Pearl Charcoal** | Thread–**Grey** | Knots–**Inside** | Seams–**Inside Floating** | Bugle Beads–**#3 Black** | Fitted Top— Fabric weight–**Medium** | Fabric color–**Black** | Thread–**Black** | Knots–**Inside** | Seams–**Inside Felled** | Binding Stitch–**Cretan P. 148 Swatch 1** Fabric weight– **Medium** | Top-layer–**Silt** | Backing–**White** | Appliqué #1–**Silt** | Appliqué #2–**White** | Pleated Ruffle–**White** | Stitch–**Cretan** | Stencil–**June's Spring** | Treatment–**Mixed** | Textile paint–**Cream** | Thread–**Tan** | Embroidery floss–**Tea** | Knots–**Inside** | Chop beads–**White** | Bugle beads–**#3 White** | Sequins–**White and Iridescent** | **Swatch 2** Fabric weight–**Medium** | Top-layer–**Silt** | Backing–**Silt** | Appliqué–**Silt** | Pleated Ruffle–**Twilight** | Stitch–**Cretan** | Thread–**Tan** | Knots–**Inside**

P. 149 Fitted Top—Fabric weight–**Medium** | Top-layer–**Twilight** | Backing– **Twilight** | Pleated Ruffle–**Black** | Appliqué–**Silt** | Stencil–**Paisley** | Treatment– **Negative Reverse Appliqué** | Textile paint–**Charcoal** | Thread–**Grey, Tan, and Black** | Knots–**Inside** | Seams–**Inside Felled** | Binding Stitch–**Cretan P. 150** Fabric weight–**Medium** | Top-layer–**Plum** | Backing–**Plum** | Couching ropes–**Plum** |

Stencil–**Anna's Garden** | Treatment–**Couching** | Textile paint–**Pearl Charcoal** | Thread–**Burgundy** | Knots–**Inside P. 151 Model 1** Sleeveless T-Shirt Top—Fabric weight–**Medium** | Fabric color–**Nude** | Thread–**Tan** | Knots–**Inside** | Seams–**Inside Felled** | Binding Stitch–**Cretan** | Short Skirt—Fabric weight–**Medium** | Top-layer– **Black** | Backing–**Black** | Stencil–**Stars** | Treatment Satin Stars | Textile paint–**Slate** | Thread–**Black** | Knots–**Inside** | Seams–**Inside Felled** | Binding Stitch–**Zigzag** | Bugle beads–**#3 Black** | Chop beads–**Black** | Sequins–**Black Model 2** Fitted Baby Doll Top—Fabric weight–**Medium** | Fabric color–**Deep** | Thread–**Grey** | Knots–**Inside** | Seams–**Inside Felled** | Binding Stitch–**Cretan** | Ruffle Stitch– **Zigzag Chain** | Short Skirt—Fabric weight–**Medium** | Top-layer–**Plum** | Backing– **Plum** | Couching ropes–**Plum** | Stencil–**Anna's Garden** | Treatment **Couching** | Textile paint–**Pearl Charcoal** | Thread–**Burgundy** | Knots–**Inside** | Seams–**Inside Felled** | Binding Stitch–**Zigzag**

P. 152 Fabric weight–**Medium** | Top-layer–**Twilight** | Backing–**Black** | Stencil– **Anna's Garden** | Treatment–**Outside Reverse Appliqué** | Textile paint– **Pearl Charcoal** | Thread–**Grey** | Knots–**Inside P. 153** Bucket Hat—Fabric weight–**Medium** | Top-layer–**Deep** | Backing–**Black** | Stencil–**Anna's Garden** | Treatment–**Outside Reverse Appliqué** | Textile paint–**Pearl Charcoal** | Thread– **Grey** | Knots–**Inside** | Seams–**Inside Floating** | Fitted Dress—Fabric weight– **Medium** | Top-layer–**Twilight** | Backing–**Black** | Stencil–**Anna's Garden** | Treatment–**Outside Reverse Appliqué** | Textile paint–**Charcoal** | Thread–**Grey** | Knots–**Inside** | Seams–**Inside Felled** | Binding Stitch–**Cretan** | Mid-Length Skirt—Fabric weight–**Light** | Fabric color–**Black** | Treatment–**Beaded Border** | Thread–**Black** | Knots–**Inside** | Seams–**Inside Felled** | Binding Stitch–**Zigzag** | Chop beads–**Black** | Bugle beads–**#3 Black** | Sequins–**Black and Dk. Brown P. 154** Bucket Hat—Fabric weight–**Medium** | Top-layer–**Black** | Backing–**Black** | Appliqué–**Black** | Stencil–**June's Spring** | Treatment–**Mixed** | Textile paint–**Slate** | Thread–**Black** | Embroidery floss–**Black** | Knots–**Inside** | Seams–**Inside Floating** | Chop beads–**Black** | Bugle beads–**#3 Black** | Sequins–**Black and Dk Brown** | Fitted Dress—Fabric weight–**Medium** | Top-layer–**Black** | Backing–**Black** | Appliqué–**Black** | Stencil–**Facets** | Treatment–**Special Appliqué** | Textile paint– **Charcoal** | Thread–**Black** | Embroidery Thread–**Black** | Knots–**Inside** | Seams– **Inside Felled** | Binding Stitch–**Cretan** | Bugle beads–**#3 Black** | Short Skirt— Fabric weight–**Medium** | Top-layer–**Twilight** | Backing–**Black** | Stencil–**Anna's Garden** | Treatment–**Outside Reverse Appliqué** | Textile paint–**Charcoal** | Thread–**Grey** | Knots–**Inside** | Seams–**Inside Felled** | Binding Stitch–**Zigzag**

P. 155 Fabric weight–**Medium** | Top-layer–**Twilight** | Backing–**Black** | Stencil–**Anna's Garden** | Treatment–**Beaded Backstitch Reverse Appliqué** | Textile paint–**Charcoal** | Thread–**Black** | Knots–**Inside** | Bugle beads–**#3 Black**
P. 156 & 157 Tied Wrap–Fabric weight–**Medium** | Top-layer–**Black** | Backing–**Black** | Stencil–**Spiral** | Treatment–**Embroidery** | Textile paint–**Charcoal** | Thread–**Black** | Embroidery floss–**Variegated** | Knots–**Outside** | Seams–**Outside Floating** | Fitted Top—Fabric weight–**Medium** | Fabric color–**Deep** | Thread–**Grey** | Knots–**Inside** | Seams–**Inside Felled** | Binding Stitch–**Cretan** | Mid-Length Skirt—Fabric weight–**Light** | Fabric color–**Black** | Thread–**Black** | Knots–**Inside** | Seams–**Inside Felled** | Binding Stitch–**Zigzag** | Long Fluted Sleeve Bolero—Fabric weight–**Light** | Fabric color–**Black** | Treatment–**Beaded Border** | Thread–**Black** | Knots–**Inside** | Seams–**Inside Felled** | Binding Stitch–**Cretan** | Chop beads–**Black** | Bugle beads–**#3 Black** | Sequins–**Black and Dk. Brown** | Fitted Top with Ruffle—Fabric weight–**Medium** | Fabric color–**Charcoal** | Ruffle–**Deep** | Appliqué #1–**Charcoal** | Appliqué #2–**Deep** | Treatment–**Beaded Appliqué Stripe** | Stitch–**Cretan** | Thread–**Grey** | Knots–**Inside** | Seams–**Inside Felled** | Binding Stitch–**Cretan** | Ruffle Stitch–**Zigzag Chain** | Chop beads–**Black** | Short Skirt—Fabric weight–**Medium** | Top-layer–**Black** | Backing–**Black** | Stencil–**Fern** | Treatment–**Embroidery** | Textile paint–**Slate** | Thread–**Black** | Knots–**Inside** | Seams–**Inside Felled** | Binding Stitch–**Zigzag** | Mid-Length Skirt—Fabric weight–**Light** | Fabric color–**Charcoal** | Treatment–**Beaded Border** | Thread–**Black and Grey** | Knots–**Inside** | Seams–**Inside Felled** | Binding Stitch–**Zigzag** | Chop beads–**Black** | Bugle beads–**#3 Black** | Sequins–**Black and Dk. Brown**

P. 158 & P. 159 Tied Wrap | Fabric weight–**Medium** | Top-layer–**Black** | Backing–**Black** | Stencil–**Spiral** | Treatment–**Embroidery** | Textile paint–**Charcoal** | Thread–**Black** | Embroidery floss–**Variegated** | Knots–**Outside** | Seams–**Outside Floating** | Fitted Top—Fabric weight–**Medium** | Fabric color–**Deep** | Appliqué–**Black** | Treatment–**Random Ruffle** | Stitch–**Straight** | Thread–**Grey and Black** | Knots–**Inside** | Seams–**Inside Felled** | Binding Stitch–**Cretan** | Fingerless Gloves—Fabric weight–**Light** | Top-layer–**Black** | Backing–**Black** | Stencil–**Anna's Garden** | Treatment–**Backstitch Reverse Appliqué** | Textile paint–**Slate** | Embroidery Thread–**Black** | Knots–**Inside** | Seams–**Inside Felled** | Short Skirt—

Fabric weight–**Medium** | Top-layer–**Black** | Backing–**Black** | Stencil–**Spiral + Vitae** | Treatment–**Embroidery** | Textile paint–**Slate and Charcoal** | Thread–**Black** | Embroidery floss–**Black** | Knots–**Outside** | Seams–**Inside Felled** | Binding Stitch–**Zigzag** | Mid-Length Skirt—Fabric weight–**Light** | Fabric color–**Black** | Thread–**Black** | Knots–**Inside** | Seams–**Inside Felled** | Binding Stitch–**Zigzag** | Bucket Hat—Fabric weight–**Medium** | Top-layer–**Black** | Backing–**Black** |Stencil–**Anna's Garden** | Treatment–**Negative Reverse Appliqué** | Textile paint–**Pearl Charcoal** | Thread–**Black** | Knots–**Inside** | Seams–**Inside Floating** | Long Crochet Necklace—Fabric weight–**Medium** | Fabric color–**Black** | Fabric Swatch—Fabric weight–**Medium** | Top-layer–**Black** | Backing–**Black** | Stencil–**Vitae and Spiral** | Treatment–**Backstitch Embroidery** | Textile paint–**Pearl Brownie and Slate** | Embroidery floss–**Black** | Knots–**Outside**

P. 160 Long Crochet Necklace—Fabric weight–**Medium** | Fabric color–**Black**
P. 160 Long Crochet Necklace—Fabric weight–**Medium** | Fabric color–**Black** | Long Crochet Necklace—Fabric weight–**Medium** | Fabric color–**Plum** | Long Crochet Necklace—Fabric weight–**Medium** | Fabric color–**Sand** | Long Crochet Necklace—Fabric weight–**Medium** | Fabric color–**Charcoal** | Fitted Top—Fabric weight–**Medium** | Top-layer–**Sand** | Backing–**Sand** | Stencil–**Anna's Garden** | Treatment–**Backstitch Reverse Appliqué** | Textile paint–**Pearl Silver** | Thread–**Tan** | Embroidery Floss–**Tea** | Knots–**Inside** | Seams–**Inside Felled** | Binding Stitch–**Cretan** | Mid-Length Skirt—Fabric weight–**Light** | Top-layer–**Black** | Backing–**Black** | Stencil–**Paisley** | Textile paint–**Pearl Grape** | Treatment–**Negative Reverse Appliqué** | Thread–**Black** | Knots–**Inside** | Seams–**Inside Felled** | Binding Stitch–**Zigzag P. 161** Fitted Top—Fabric weight–**Medium** | Fabric color–**Deep** | Thread–**Grey** | Knots–**Inside** | Seams–**Inside Felled** | Binding Stitch–**Cretan** | Fingerless Gloves—Fabric weight–**Light** | Top-layer–**Black** | Backing–**Black** | Stencil–**Anna's Garden** | Treatment–**Backstitch Reverse Appliqué** | Textile paint–**Slate** | Embroidery Thread–**Black** | Knots–**Inside** | Seams–**Inside Felled P. 161** Fabric weight–**Medium** | Top-layer–**Deep** | Backing–**Deep** | Stencil–**Anna's Garden** | Treatment–**Backstitch Reverse Appliqué** | Textile paint–**Pearl Charcoal** | Embroidery floss–**Dk. Grey** | Knots–**Inside**

P. 162 & P. 163 Tied Wrap—Fabric weight–**Medium** | Top-layer–**Black** | Backing–**Black** | Stencil–**Facets** | Treatment–**Natalie's Dream** | Textile paint–

Slate | Thread–**Black** | Embroidery floss–**Black** | Knots–**Inside** | Seams–**Outside Floating** | Bugle Beads–**#3 Black** | Fitted Top—Fabric weight–**Medium** | Top-layer–**Black** | Backing–**Black** | Stencil–**Anna's Garden** | Treatment–**Negative Reverse Appliqué** | Textile paint–**Slate** | Thread–**Black** | Knots–**Outside** | Seams–**Inside Felled** | Binding Stitch–**Cretan** | Short Skirt—Fabric weight–**Medium** | Top-layer–**Black** | Backing–**Black** | Appliqué–**Black** | Treatment–**Beaded Ruffle Appliqué** | Thread–**Black** | Knots–**Inside** | Seams–**Inside Felled** | Binding Stitch–**Zigzag** | Chop Beads–**Black** | Long Skirt—Fabric weight–**Medium** | Top-layer–**Black** | Backing–**Black** | Stencil–**Anna's Garden** | Treatment–**Negative Reverse Appliqué** | Textile paint–**Slate** | Thread–**Black** | Knots–**Outside** | Seams–**Inside Felled** | Binding Stitch–**Zigzag** | Fingerless Gloves—Fabric weight–**Light** | Top-layer–**Black** | Backing–**Black** | Stencil–**Anna's Garden** | Treatment–**Backstitch Reverse Appliqué** | Textile paint–**Slate** | Embroidery Thread–**Black** | Knots–**Inside** | Seams–**Inside Felled** | Fabric Swatch—Fabric weight–**Medium** | Top-layer–**Black** | Backing–**Black** | Stencil–**Anna's Garden** | Treatment–**Negative Reverse Appliqué** | Textile paint–**Slate** | Thread–**Black** | Knots–**Outside**

First and foremost, thank you to our readers, clients, patrons, and supporters. Without you, this book—and the last decade of Alabama Chanin—would not have been possible.

And to so many, many folks, in no particular order, who have touched my life in immeasurable ways over the last decade:

Thank you to my everyday family: Billy Smith, Sherry Dean Smith, Butch Anthony, Maggie Anthony-Chanin, Zach Chanin, Myra and Jim Brown, and my grandparents, Stanley and Lucille Perkins and Aaron and Christine Smith, who nurtured the garden upon which this work is built.

And to my work family: Steven Smith, Diane Hall, Kay Woehle, Staci Phillips, Rachel Wallace, Olivia Sherif, June Flowers-Sledman, Conrad Pitts, Terry Wiley, Lisa Patterson, and the multitude of artisans who work with us.

Our extended family, who have all touched this work: Robert Rausch, Chris Timmons, Sun Young Park, Sara Martin, Briana Knight, Peter Stanglmayr, Sarah Ellison Lewis, Marisa Keris, Angie Mosier, John T. Edge, Roland McKinney, Michelle Nichols, Eva Whitechapel (for lots of advice), Elizabeth De Ramus, Sissi Farassat, Igor Orovac, and all the folks @ Fish Film (RIP), Lisa and Jess Morphew, Paul Graves and former partners, Paul Kelly and Paul McKevitt for a lovely beginning, Elaine Poorman, George Perkins, Joy Kelley, Erin Dempsey, Jessica Turner, Mrs. Jessie Mangrum, Jennifer Rausch, Tom Hendrix, Fritz Woehle, Birgit Buertlmair, the Council of Fashion Designers of America, *Vogue* magazine, Sally Singer, and Bureau Friends, Julie Gilhart, Cathy Bailey, Maria Moyer, and Nicole Mackinlay Hahn. To Alina Zakaite, Katia Inamo, Aysche Tiefenbrunner, Maria Flavia, Ivory Rose Strzalkowski, Jane Moseley, and Camille Mervin Leroy for their lovely faces and for their help in making this beautiful book. And to Veronika Dirnhofer—who first listened to me ramble on about T-shirts.

Melanie Falick, our editor, carried a luminous smile in the face of this challenge…

I am grateful to one and all.

About the Author

Natalie Chanin, former costume designer and fashion stylist, is the founder and head designer of Alabama Chanin. Her work has been featured in *Vogue*, *Time*, the *New York Times*, and *Town & Country*, as well as on CBS news. She is the author of *Alabama Stitch Book* (Abrams, 2007) and *Alabama Studio Style* (Abrams, 2009). Natalie is a member of the Council of Fashion Designers of America, and her work was selected for the 2010 Global Triennial—Why Design Now?—by the Cooper-Hewitt, National Design Museum. She works from her hometown of Florence, Alabama, as an entrepreneur, designer, writer, collector of stories, filmmaker, mother, gardener, and cook.

Most of the materials and tools called for in the projects in this book are available
from fabric and craft retailers and from the Alabama Chanin website. Kits for many
of the projects are also available on the website, www.alabamachanin.com.

Editor: **Melanie Falick**
Technical Editor: **Chris Timmons**
Designer: **Robert Rausch**
Production Manager: **Tina Cameron**

Library of Congress Cataloging-in-Publication Data

Chanin, Natalie.
  Alabama studio sewing + design : a guide to hand-sewing an Alabama Chanin
wardrobe / Natalie Chanin ; photographs by Rinne Allen ... [et al.] ;
illustrations by Sun Young Park.
      p. cm.
  Includes index.
  ISBN 978-1-58479-920-7
  1. Tailoring—Patterns. 2. Natural products—Alabama. I. Title.
  TT590.C45 2012
  687'.04409761—dc23
                        2011021853

The text of this book was composed in Arial and Palatino.

Published in 2012 by Abrams

Printed and bound in China
10 9 8 7

Abrams books are available at special discounts when purchased
in quantity for premiums and promotions as well as fundraising
or educational use. Special editions can also be created to specification.
For details, contact specialsales@abramsbooks.com or the address below.

ABRAMS The Art of Books
115 West 18th Street, New York, NY 10011
abramsbooks.com